Organic Chemistry

A Guide to Common Themes

Organic Chemistry

A Guide to Common Themes

Trevor M. Kitson

Department of Chemistry & Biochemistry
Massey University
Palmerston North
New Zealand

Edward Arnold
A division of Hodder & Stoughton
LONDON MELBOURNE AUCKLAND

First published in Great Britain 1988
Reprinted 1991

British Library Cataloguing in Publication Data

Kitson, Trevor M.
Organic chemistry: a guide to common themes
1. Chemistry, Organic.
I. Title
547 QD251.2
ISBN 0 7131 3649-9

Typeset in 10/12 Times Linoterm by Keyset Composition, Colchester.
Printed and bound in Great Britain for Edward Arnold,
a division of Hodder and Stoughton Limited,
Mill Road, Dunton Green, Sevenoaks, Kent TN13 2YA
by Athenaeum Press Ltd, Newcastle upon Tyne.

Preface

The purpose of this book is to bring to the attention of university students certain threads that run throughout all branches of organic chemistry and that result in 'connections' amongst various seemingly separate areas.

In most university lecture courses, the limited time available often means that students and teachers have to deal with the subject from a more or less **linear**, 'string of beads' approach. The first bead might be alkanes, the second alkenes, then alkyl halides, alcohols, and so on. However, there are many more points of contact between these 'beads' than this straightforward linear approach would suggest. Rather than being just one simple necklace, organic chemistry is more like a complex knotted tangle of jewellery. Or to use a chemical analogy, the subject is not a linear polymer, but a highly **cross-linked** one.

Thus rather than approach organic chemistry from the point of view of different **classes of compound** (as all of the popular standard texts do), I have collected together and summarised in this book observations on different **types of intermediate** (e.g. all reactions involving carbanions), different **types of reaction** (e.g. all those exemplifying electrophilic attack) and different useful **concepts** (e.g. all those points illuminated by the idea of resonance). All these various topics cut right across the usual boundaries set by classes of compound—for instance, carbocations are met in the reactions of alkenes, alcohols, amines, arenes, alkyl halides, etc., as we will see.

The idea, therefore, is for students first to acquire some straightforward factual knowledge of organic chemistry—names, structures, properties, reactions, syntheses, etc.—and then to use this book to re-examine and re-classify what they have learnt. In this way they will recognise the common unifying themes that run through what may have appeared at first sight to be an immense and chaotic subject.

It is my belief that looking at 'connections' in this way constitutes a small investment of hard work that pays out a huge dividend in interest, understanding, enjoyment, and simplification of the subject.

Pushing arrows

It is vitally important that you become familiar with organic chemistry's curly arrows. An arrow symbolises the movement of an **electron-pair**. Now when electrons move, the atoms involved gain or lose positive or negative charges, and you must keep account of the charges when you are pushing arrows. Sometimes a student is confused by the appearance of **single** charges when an electron-**pair** moves. Let us consider an example in detail and see how this happens:

$$R{-}\ddot{N}H_2 \quad R_2C{=}O \rightarrow R{-}\overset{+}{N}H_2{-}CR_2{-}O^-$$

(This might be the first step in the familiar reaction of 2,4-dinitrophenylhydrazine with propanone.) First of all, revise atomic structure until you are quite happy that the species RNH_2 should be neutral to start with and that the nitrogen carries a lone pair of electrons. Now the arrow shows the lone-pair of the nitrogen moving to form a new bond to carbon, but the nitrogen has not **lost** both these electrons—it still **shares** them (with carbon). It used to own them both entirely; now it shares them—consequently it has effectively lost a single negative charge and we therefore write a positive sign on the nitrogen. In more detail:

$$N{:} \quad C \vdots \ddot{O}{:} \rightarrow N^+ \div C \div \ddot{O}{:}^-$$

[A more fundamental way of looking at it is to say that the nitrogen has *two* electrons in its inner ($1s^2$) shell, and a half-share in each of the electron-pairs in its *four* covalent bonds, making an effective total of *six* electrons. In the nucleus, however, there are *seven* protons—hence one net positive charge.] Consider the oxygen in the same way. The arrow shows movement of one of the electron-pairs in the carbonyl double bond on to the oxygen. Originally the oxygen **shared** this pair of electrons (with carbon); later it possesses them all to itself—effectively therefore it has gained a single negative charge (and that is why we now write it as $-O^-$). Overall the product must be neutral of course as both starting materials were neutral.

Until arrow pushing becomes second nature, you should try keeping track of every electron and charge in this way for the various reaction mechanisms you will meet below.

Contents

1

Carbocations

1.1 Introduction

A carbocation (or carbonium ion as it used to be known) can be thought of formally as being derived from an alkane by removal of a hydride ion (although this is not a practical method of preparation).
e.g.

$$\mathrm{H\!:\!\overset{\displaystyle H}{\underset{\displaystyle H}{\overset{\cdot|\cdot}{\underset{\cdot|\cdot}{C}}}}\!:\!H \rightarrow H\!:\!\overset{\displaystyle H}{\underset{\displaystyle H}{\overset{\cdot|\cdot}{\underset{\cdot|\cdot}{C^{+}}}}} + :H^{-}}$$

The species which results has only six electrons in its valence shell and is therefore very reactive. It is also **planar** and trigonal in shape, with sp^2-hybridised bonds, so it would be better represented as

$$\mathrm{H\!:\!\overset{+}{C}\!\begin{smallmatrix} \diagup H \\ \diagdown H \end{smallmatrix}}$$

Later we will meet carbocations which are highly resonance stabilised (such as the triphenylmethyl and tropylium ions); consequently these species are relatively unreactive and long lived. It is important to realise however, that ordinary simple carbocations (such as CH_3^+ above) are very reactive indeed and have only a fleeting existence as intermediates in certain reactions.

1.2 Reactions of carbocations

There are three processes which this very reactive species might undergo. (1) It might **combine with a nucleophile**, which would not be surprising as the carbocation is positively charged, and a nucleophile by definition is something which 'loves' positive charges. By the same token, the carbocation is an electrophile. (2) It might **lose a proton**, to give a neutral unsaturated compound (an alkene). (3) It might **rearrange** first and then do either (1) or (2).

The diazotisation of a simple aliphatic primary amine illustrates well the occurrence of these three processes. See Fig. 1.1. The products obtained from 1-butanamine are shown with their yields in parenthesis. Obviously this is a synthetically useless reaction. Many other reactions involving carbocations give

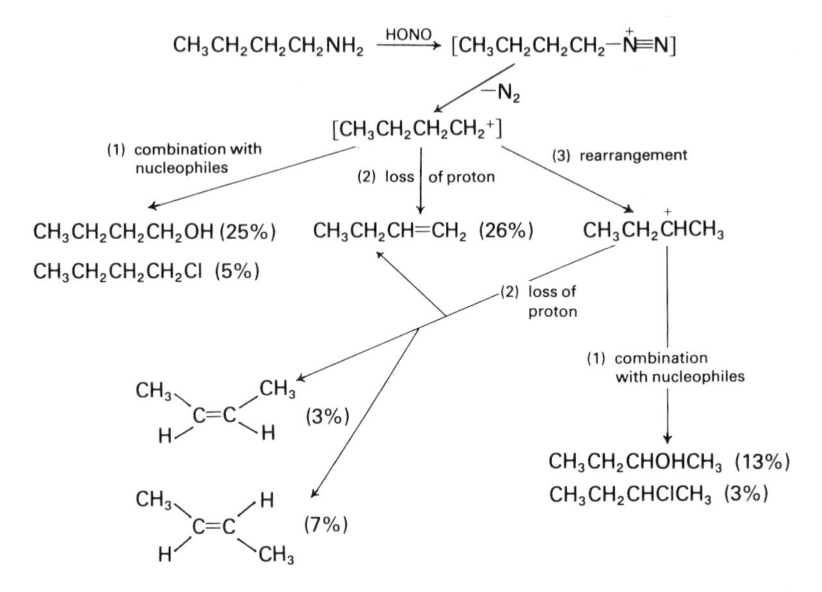

Fig. 1.1

a particular product in much better yield but even so we must always be aware of the possibilities for side reactions. In particular, as we shall see, the potential for rearrangement [process (3)], can often lead to unexpected results. (But *not* unexpected to the experienced chemist!)

Let us now look at some of the more familiar reactions involving carbocations, and see how they exemplify our three key processes.

(1) Combination with a nucleophile
(2) Loss of a proton
(3) Rearrangement

(a) Substitution reactions of alkyl halides. The S_N1 reaction

[See Sections 5.3 and 11.3] Certain alkyl halides spontaneously ionise to some extent and the resulting carbocation can then react with a nucleophile [*process (1)*]

e.g.

$$(CH_3)_3C{-}Br \rightleftharpoons (CH_3)_3C^+ + Br^-$$

$$\downarrow H_2O$$

$$(CH_3)_3C{-}\overset{+}{O}H_2 \xrightarrow{-H^+} (CH_3)_3C{-}OH$$

The first step is the slow, rate-determining step and so the rate of reaction is independent of the nucleophile concentration (in this example, water). As we noted above, the carbocation is planar and this means that the nucleophile can become attached equally readily from one side or the other, leading to **racemisation** (in cases where the original alkyl halide was chiral).

e.g.

$$\text{Me(Et)(Pr)C–Br} \underset{}{\overset{-\text{Br}^-}{\rightleftharpoons}} \text{Et(Pr)C}^+\text{–Me} \xrightarrow{:\text{OH}_2} \text{Et(Pr)(Me)C–}\overset{+}{\text{O}}\text{H}_2 \text{ (both configurations)}$$

(Me = methyl, Et = ethyl, Pr = propyl)

(loss of H^+ gives the alcohol)

A reaction which often competes with an S_N1 reaction is that known as E1. In this, the carbocation gives up a proton (to a base such as water) and forms an alkene [*process (2)*].
e.g.

$$(CH_3)_3C^+ \xrightarrow{-H^+} (CH_3)_2C{=}CH_2$$

(b) Addition of H—X to alkenes

[See also section 4.3] The addition of electrophiles such as sulphuric acid and hydrogen halides to alkenes involves as a first step protonation to give a carbocation, and then combination of the carbocation with a nucleophile [*process (1)*].
e.g.

$$CH_2{=}CH_2 + H_2SO_4 \rightarrow CH_3CH_2^+ + HSO_4^- \rightarrow CH_3CH_2{-}O{-}SO_2{-}OH$$

The resulting alkyl hydrogen sulphate may be hydrolysed to an alcohol. Alternatively, in acid-catalysed hydration, the carbocation may combine directly with water as the nucleophile:
e.g.

$$CH_3CH_2^+ + H_2O \rightarrow CH_3CH_2\overset{+}{O}H_2 \xrightarrow{-H^+} CH_3CH_2OH$$

An example with hydrogen halide:

$$(CH_3)_2C{=}CH_2 + HCl \rightarrow (CH_3)_2C^+{-}CH_3 + Cl^- \rightarrow (CH_3)_3C{-}Cl$$

(c) Dehydration of alcohols

[See also section 8.2(c)] This exemplifies *process (2)*, loss of a proton from the carbocation (i.e. an E1 mechanism).

e.g.

$$\underset{\text{OH}}{CH_3CHCH_3} \xrightarrow[\text{heat}]{H^+} \underset{^+OH_2}{CH_3CHCH_3} \xrightarrow{-H_2O} CH_3\overset{+}{C}HCH_3 \xrightarrow{-H^+} CH_3CH{=}CH_2$$

(The proton is lost to a base such as water.)

(d) The Friedel–Crafts alkylation reaction

[See also sections 4.4(d), 6.3(b)] An alkyl halide and $AlCl_3$ form a complex together as follows:

$$R{-}Cl + AlCl_3 \rightarrow R{-}Cl{\cdots}AlCl_3 \leftrightarrow R^+ \; AlCl_4^-$$

For practical purposes we can consider the reactive species to be a carbocation, R^+, although it is not completely dissociated from the $AlCl_4^-$. In the Friedel–Crafts reaction this then attacks an aromatic compound such as benzene—another example of *process (1)*, since here benzene is acting as a nucleophile (and R^+ as an electrophile) in just the same way as OH^-, HSO_4^- and Cl^- have been nucleophiles in the examples we have looked at.

$$C_6H_6 + R^+ \longrightarrow [C_6H_6R]^+ \longleftrightarrow \text{other resonance forms [see section 10.7(a)]}$$

In the second step the intermediate (which is in fact another carbocation) loses a proton [*process (2)* again] to regenerate the stable aromatic benzene ring:

$$[C_6H_6R]^+ \xrightarrow{-H^+} C_6H_5R$$

1.3 Order of stability of carbocations

Let us now look back at these reactions involving carbocations and consider some observations which give us a clue to the relative stabilities of primary, secondary and tertiary (1°, 2° and 3°) carbocations.

To start with, the S_N1 type of reaction we have seen above for *t*-butyl bromide (2-bromo-2-methylpropane) is not characteristic of *n*-butyl bromide (1-bromobutane). In other words, 1-bromobutane has a negligible tendency to ionise, suggesting that the 1° carbocation which would be formed is of very high energy—that is, less stable than the 3° carbocation which *is* formed in the reactions of the 3° alkyl halide:

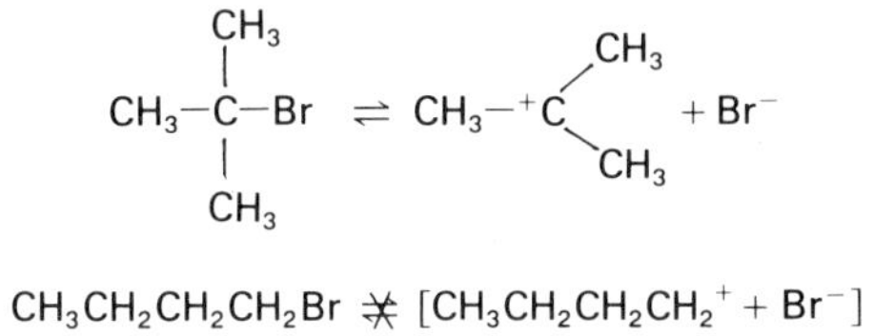

$$(CH_3)_3C{-}Br \rightleftharpoons CH_3{-}{}^{+}C(CH_3)_2 + Br^-$$

$$CH_3CH_2CH_2CH_2Br \not\rightleftharpoons [CH_3CH_2CH_2CH_2^{+} + Br^-]$$

Secondly we have seen above that methylpropene reacts with HCl to give 2-chloro-2-methylpropane, not the alternative product (if H and Cl added the other way round) 1-chloro-2-methylpropane. This shows that the 3° carbocation is formed preferentially over the 1° one, suggesting again that the 3° one is the more stable of the two. (The outcome of this addition reaction is of course *predicted*, but not *explained*, by Markovnikov's rule.)

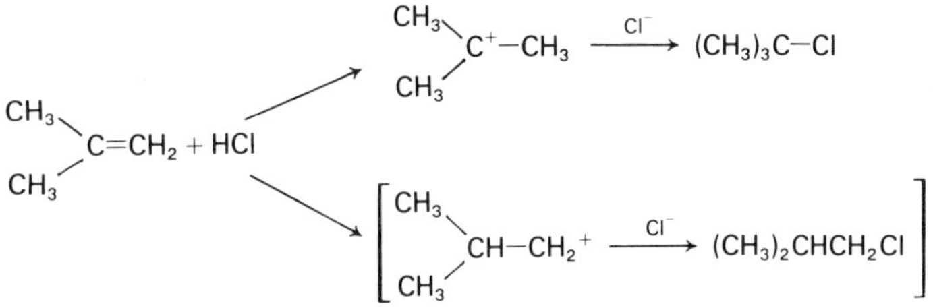

$$(CH_3)_2C{=}CH_2 + HCl \longrightarrow (CH_3)_2C^{+}{-}CH_3 \xrightarrow{Cl^-} (CH_3)_3C{-}Cl$$

$$(CH_3)_2C{=}CH_2 + HCl \longrightarrow \left[(CH_3)_2CH{-}CH_2^{+} \xrightarrow{Cl^-} (CH_3)_2CHCH_2Cl\right]$$

Thirdly, some alcohols are dehydrated by H_2SO_4 more easily than others. e.g.

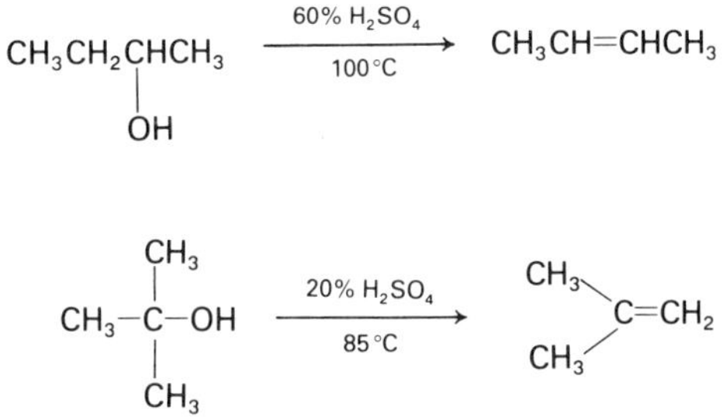

$$CH_3CH_2CH(OH)CH_3 \xrightarrow[100\,°C]{60\%\ H_2SO_4} CH_3CH{=}CHCH_3$$

$$(CH_3)_3C{-}OH \xrightarrow[85\,°C]{20\%\ H_2SO_4} (CH_3)_2C{=}CH_2$$

Once again this suggests that the 3° carbocation formed during the latter reaction is relatively stable, by which I mean relative to the 2° one which is formed in the dehydration of 2-butanol.

These examples, and many others, lead us to conclude that this is the order of stability of carbocations:

$$3° > 2° > 1°$$

There are some exceptions to this order of stability. We will see shortly several examples of carbocations which appear at first sight to be primary, but which, because of resonance stabilisation, are formed very much more readily than might be expected. On the other hand, tertiary carbocations which cannot take up a planar shape are hardly every formed. For example, contrast the following:

$CH(CH_2CH_2)_3C{-}Br$

ionises 1 million times less readily than *t*-butyl bromide; the three alkyl groups are tied back and cannot become planar

$(CH_3)_3C{-}Br$

t-butyl bromide; ionises fairly readily

1.4 Rearrangements of carbocations

Perhaps the most convincing evidence for the general stability order for carbocations (3° > 2° > 1°) is that rearrangements often occur in reactions involving carbocations, and the rearrangements always occur in the same direction—that is, 1° carbocations rearrange to give 2° or 3° ones, and 2° ones give 3°, but not vice versa. Here is an example:

$$CH_3CH_2CH_2CH_2OH \xrightarrow{H_2SO_4} CH_3CH{=}CHCH_3 \text{ (more than } CH_3CH_2CH{=}CH_2)$$

This involves:

$$CH_3{-}CH_2{-}CH_2{-}\overset{+}{C}H_2 \xrightarrow[\text{hydride ion}]{\text{shift of a}} CH_3{-}CH_2{-}\overset{+}{C}H{-}CH_3$$

less stable 1° — more stable 2°

loss of proton

$$CH_3CH{=}CHCH_3$$

And a second example:

$$C_6H_6 + (CH_3)_2CH{-}CH_2Cl \xrightarrow{AlCl_3} C_6H_5{-}C(CH_3)_2{-}CH_3 \quad (\text{not } C_6H_5{-}CH_2CH(CH_3)_2)$$

This involves:

$$CH_3{-}\overset{H}{\underset{CH_3}{C}}{-}\overset{+}{C}H_2 \longrightarrow CH_3{-}\underset{CH_3}{\overset{+}{C}}{-}CH_3$$

less stable 1° more stable 3°

1.5 Make connections!

At this stage let us pause and think again of the purpose this book is designed for. It is to draw out and emphasise the similarities in what at first sight may seem completely different and separate reactions. In fact we have some very good examples here—reactions as diverse as the hydrolysis of *t*-butyl bromide, the addition of HI to 1-butene, the substitution of an ethyl group into benzene, etc., are all *connected* because they all involve the intermediacy of carbocations. Everything you know about acid-catalysed dehydration of alcohols can be connected to everything you know about Markovnikov's rule—because behind it all is the relative order of stability of carbocations. The unsuccessful student is the one who will laboriously learn that certain alcohols undergo rearrangement when dehydrated and also that certain alkyl groups rearrange during the Friedel–Crafts reaction—and not make the connection. The successful student will see that both these reactions involve *exactly the same process*. Furthermore, if she meets some new reaction also involving carbocations she will know what sort of points to look out for (e.g. relative ease of reaction of 1°, 2° and 3° compounds, rearrangements, and so on). And if she ever gets to the lofty heights of carrying out research into a reaction of *unknown* mechanism she may make various experimental observations and be able to say 'Eureka! These are exactly the characteristics I would expect for a reaction involving a carbocation.'

1.6 More on the Friedel–Crafts reaction

A good example of the theme that I have just been discussing is found in the way the Friedel–Crafts reaction has been extended. This reaction was originally developed with alkyl halides and $AlCl_3$ (as catalyst) and that is still how it is usually carried out in the laboratory. However, as you have seen we can 'connect' this reaction to the dehydration of alcohols [section 1.2(c)] and to the hydration of alkenes [section 1.2(b)] since they all involve carbocations. Knowing this enabled industrial chemists to use alcohols or alkenes (with suitable acid catalysts) as starting materials for the Friedel–Crafts reaction. (Simple alcohols and alkenes such as 2-propanol and propene are comparatively cheap—an important point in industry.)

$$CH_3{-}CHCl{-}CH_3 + AlCl_3 \longrightarrow CH_3\overset{+}{C}H\,CH_3$$

$$CH_3{-}CHOH{-}CH_3 \xrightarrow{H^+} CH_3{-}CH\overset{+}{O}H_2{-}CH_3 \xrightarrow{-H_2O} CH_3\overset{+}{C}H\,CH_3$$

$$CH_3{-}CH{=}CH_2 + H^+ \longrightarrow CH_3\overset{+}{C}H\,CH_3$$

(all the starting materials give this common intermediate)

1.7 More reactions involving carbocations

In this section we will look briefly at some more reactions involving carbocations to see how they 'connect' with what has been discussed already.

(a) Hydrolysis of *t*-butyl esters

The usual mechanism of hydrolysis of esters will be covered later [sections 5.5(l), 5.5(n)]; at this stage we are interested in a mechanism which only happens with esters having a 3° group such as in the following:

$$(CH_3)_3C{-}O{-}\overset{O}{\overset{\|}{C}}{-}CH_3 \overset{H^+}{\rightleftharpoons} (CH_3)_3C{-}O{-}\overset{\overset{+}{O}H}{\overset{\|}{C}}{-}CH_3 \rightarrow (CH_3)_3C^+ + O{=}\overset{OH}{\overset{|}{C}}{-}CH_3$$

$$(CH_3)_3C^+ \xrightarrow{H_2O} (CH_3)_3C{-}\overset{+}{O}H_2 \xrightarrow{-H^+} (CH_3)_3C{-}OH$$

The key step is dissociation of the protonated ester to give the carboxylic acid and the carbocation; it is not surprising from what has already been said (about, for example, S_N1 reactions, the ease of dehydration of *t*-butyl alcohol, the general order of stability of carbocations, etc.) that this mechanism only occurs when a **tertiary** carbocation is formed. It would not happen with, for example, $CH_3{-}OCOCH_3$ (methyl ethanoate).

Evidence to support the mechanism shown above is that hydrolysis of *t*-butyl acetate (dimethylethyl ethanoate) in the presence of $H_2{}^{18}O$ gives ^{18}O-labelled *t*-butyl alcohol but *un*labelled acetic acid.

(b) The pinacol rearrangement

Consider this reaction mechanism:

$$(CH_3)_2C(OH)-C(OH)(CH_3)_2 \xrightarrow{H^+} (CH_3)_2C(OH)-C(^+OH_2)(CH_3)_2 \xrightarrow{-H_2O} CH_3-C(CH_3)(OH)-C^+(CH_3)_2$$

dimethyl-2,3-butanediol
(pinacol)

$$\downarrow$$

$$\left\{ CH_3-\overset{+}{C}(:OH)-C(CH_3)_2-CH_3 \longleftrightarrow CH_3-C(={}^+OH)-C(CH_3)_2-CH_3 \right\}$$

$$CH_3-C(=O)-C(CH_3)_3 \xleftarrow{-H^+}$$

dimethylbutanone
(pinacolone)

In section 1.4 we saw examples of the rearrangement of 1° carbocations to more stable 2° or 3° ones by migration of a **hydride ion** (i.e. hydrogen with its bonding pair of electrons). Now the pinacol rearrangement above (which is a classic reaction known since 1860) is rather similar in that it too involves the rearrangement of a carbocation, but this time by means of the migration of a **methyl group** with its bonding pair of electrons. But, you might ask, why does it migrate since in this case the first formed carbocation is already tertiary? The answer is that, after migration, the carbocation is **stabilised by resonance** as drawn above. In fact the lower of the two resonance structures is by far the more important contributor to the hybrid as here all the atoms have 'inert gas' configurations—unlike a true carbocation where carbon has only six electrons in its valence shell. (This species is really just a protonated ketone.)

The point I am making here is that the driving force of the reaction is *rearrangement of a carbocation to a more stable structure*—exactly the same kind of thing we have seen in dehydration of some alcohols, and in some Friedel–Crafts reactions.

(c) Hydroboration

[See also sections 4.3(h) and 11.7(d)] We have explained the operation of Markovnikov's rule in the addition of H—X to alkenes in terms of the order of stability of carbocations (or to put it a better way, the fact that many reactions follow Markovnikov's rule gives us a strong clue as to the relative stability of carbocations). Hydroboration followed by oxidation results in *anti*-Markovnikov hydration of an alkene.

e.g.

$$CH_3CH{=}CH_2 \xrightarrow[\text{(ii) } H_2O_2/OH^-]{\text{(i) } B_2H_6} CH_3CH_2CH_2OH$$

Does this mean that hydroboration has nothing to do with the stability of carbocations? Not necessarily! Consider the following, which is a plausible reaction mechanism. (The reagent diborane is really B_2H_6 and the nature of the bonding that it possesses makes fascinating reading. However, it is easier to follow the reaction of diborane with alkenes if we think of it as BH_3.)

$$>C{=}C< \;\; H{-}BH_2 \rightarrow -C^{\delta+}{=}{=}{=}C- \;\; H \cdots {}^{\delta-}BH_2 \rightarrow -C(H)-C(BH_2)-$$

This mechanism probably never involves a full blown carbocation (because rearrangements are *not* observed), but during the progress of the reaction a partial positive charge starts to form on one carbon of the alkene bond as electrons drift from it to become bonded to the boron atom. At more or less the same time a hydride ion is transferred to this incipient carbocation. Now consider BH_3 adding to an unsymmetrical alkene such as propene. Which of the following modes would be preferred?

$$CH_3{-}\overset{\delta+}{CH}{=}{=}{=}CH_2 \;\; (CH_2 \cdots {}^{\delta-}BH_2{-}H) \qquad CH_3{-}CH{=}{=}{=}\overset{\delta+}{CH_2} \;\; (CH \cdots {}^{\delta-}BH_2{-}H)$$

The answer is the first one, since this resembles a 2° carbocation, whereas the second one is like a 1° carbocation. (From the first mode of addition we eventually arrive at 1-propanol, whereas acid-catalysed hydration of propene would of course give 2-propanol.)

Clearly there is a major difference between hydroboration and many other additions to alkenes—in the former, hydrogen adds essentially as a hydride ion, whereas from H—Cl, H_2SO_4, H_3O^+, etc., it adds as a proton. Nevertheless, the outcome of all these reactions can be rationalised in terms of carbocation stability.

It should be pointed out, however, that the explanation above is not the only one for the anti-Markovnikov course of hydroboration. An explanation based on steric factors is also available. According to this, carbon-1 of propene is simply more sterically accessible to BH_3 than is carbon-2 (which carries the methyl group). This factor would be more significant as the reaction progresses, when RBH_2 and then R_2BH add to the second and third molecule of propene respectively. (Finally, R_3B is converted to 3ROH by treatment with alkaline hydrogen peroxide.)

(d) Acid-catalysed polymerisation of alkenes

We have seen that an alkene can combine with a proton to give a carbocation. [See sections 1.2(b), 1.6].

e.g.

$$(CH_3)_2C{=}CH_2 + H^+ \rightarrow (CH_3)_2C^+{-}CH_3$$

A recurring theme has been that this carbocation can then add to a nucleophile (i.e. something which can furnish the electrons the carbocation lacks) such as H_2O, Cl^-, HSO_4^-, benzene, etc. Another molecule of the same alkene can also provide electrons:

$$CH_3{-}C^+(CH_3)_2 \leftarrow CH_2{=}C(CH_3)_2 \rightarrow CH_3{-}C(CH_3)_2{-}CH_2{-}C^+(CH_3)_2$$

This is therefore another example of *process (1)* the combination of a carbocation with a nucleophile. In this case the product is another carbocation which can then add another molecule of alkene again and again leading eventually to a **polymer**.

1.8 Basis of the relative stability of carbocations

So far we have treated the relative stabilities of carbocations on a purely empirical basis—that is, we have noted that many experimental observations lead us to conclude that 3° carbocations are more stable than 2° carbocations which in turn are more stable than 1° carbocations. Now we ask the question: Why? It is a fundamental conclusion from physics that the more an electrical charge is dispersed (spread over a larger area) the more stable the system becomes. Thus any factor which can disperse the positive charge on a carbocation will stabilise it. It would appear therefore that, for example, the three methyl groups in the *t*-butyl cation can disperse the positive charge more than the three hydrogens can in the methyl cation since the former is observed experimentally to be much more stable than the latter. We express this in terms of the **inductive effect** of alkyl groups, which is their tendency to 'push' electrons towards an adjacent positive charge, and we represent it by an arrowhead on the bond:

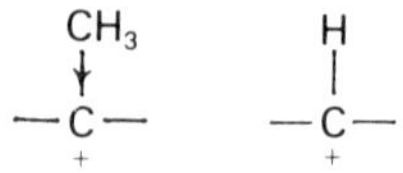

$$\overset{CH_3}{\downarrow}\quad\quad \overset{H}{|}$$
$$-\underset{+}{C}-\quad\quad -\underset{+}{C}-$$

$-CH_3$ 'pushes' electrons more than $-H$ does

The more alkyl groups around the central carbon the greater the stabilisation; hence the stability order 3° > 2° > 1° is explained. Of course, the positive charge is not eliminated, it is merely spread out a bit. We could perhaps represent the *t*-butyl cation as:

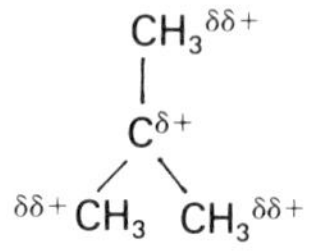

$$CH_3^{\delta\delta+}$$
$$|$$
$$C^{\delta+}$$
$$/ \quad \backslash$$
$$^{\delta\delta+}CH_3 \quad CH_3^{\delta\delta+}$$

Where $\delta+$ means somewhat less than a full positive charge and $\delta\delta+$ means a very small fraction of a full positive charge. The bonding electrons have shifted slightly towards the central carbon leaving the three methyl groups with this slight positive charge.

An alternative way of looking at the stability of a tertiary carbocation is provided by the concept of hyperconjugation [see section 10.10(a)].

1.9 Resonance stabilisation of carbocations

[See also section 10.10] Apart from the inductive effect, the other main way in which a positive charge on a carbocation can be stabilised is by **resonance**. Consider the following ionisation energies:

$$CH_3CH_2Cl \rightarrow CH_3CH_2^+ + Cl^- \qquad \Delta H = 830\ kJ\ mol^{-1}$$

$$CH_2{=}CH{-}CH_2Cl \rightarrow CH_2{=}CH{-}CH_2^+ + Cl^- \qquad \Delta H = 720\ kJ\ mol^{-1}$$

$$Ph{-}CH_2{-}Cl \rightarrow Ph{-}CH_2^+ + Cl^- \qquad \Delta H = 690\ kJ\ mol^{-1}$$

(Ph = [benzene ring]—, the phenyl group. $Ph{-}CH_2{-}$ = the benzyl group.)

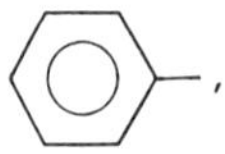

Clearly the **allyl** and **benzyl** carbocations ($CH_2{=}CH{-}CH_2^+$ and $Ph{-}CH_2^+$, respectively) are easier to form (that is, more stable) than a simple 1° carbocation even though they themselves seem to have the positive charge residing on a 1° carbon atom. The explanation is that this positive charge is not **localised** at this position but rather **delocalised** as we can show by writing the following resonance structures:

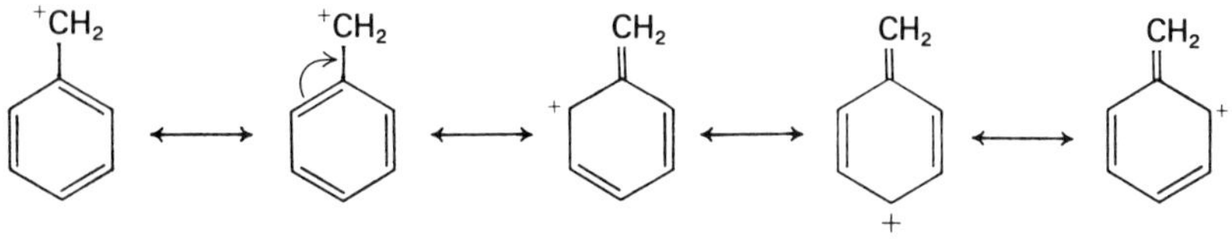

$$CH_2{=}CH{-}\overset{+}{C}H_2 \longleftrightarrow \overset{+}{C}H_2{-}CH_2{=}CH_2$$

Thus in each case the positive charge is dispersed over the ion rather than concentrated at a single point, and the ions are consequently about as stable as a simple 3° one rather than a simple 1° one as might be otherwise expected.

A kind of resonance-stabilised carbocation rather similar to the benzyl case is that which occurs in electrophilic aromatic substitution.

e.g.

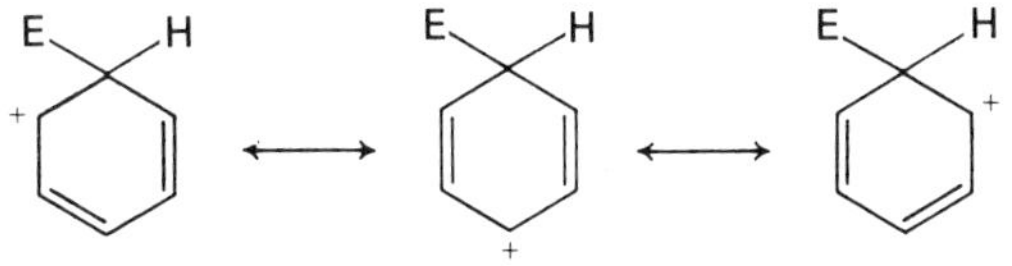

(E = an electrophile)

[We will deal with this more fully in section 10.7.]

The type of delocalisation we have just seen above for the benzyl carbocation becomes more important the more phenyl groups are present, culminating in the relatively very stable triphenylmethyl carbocation:

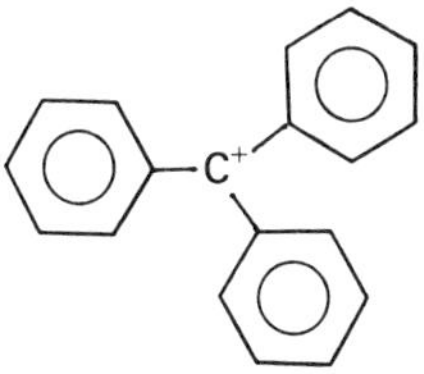

How many resonance structures can you draw for this?

This ion is so stable that when triphenylmethyl chloride is dissolved in an inert solvent such as liquid sulphur dioxide it dissociates into Ph_3C^+ and Cl^- to such an extent that the solution conducts electricity.

One of the most stable carbocations is the tropylium (or 1,3,5-cycloheptatrienyl) ion, formed for example when 7-bromo-1,3,5-cycloheptatriene ionises in water:

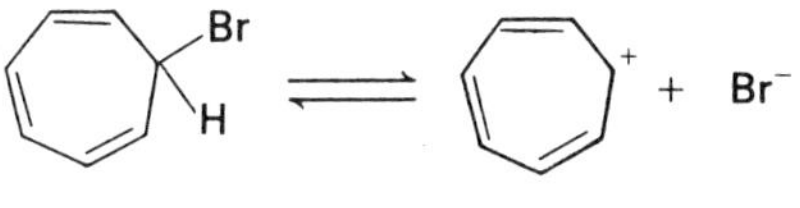

$$\rightleftharpoons \quad + \quad Br^-$$

We can draw seven equivalent resonance structures for this ion (can you?) showing that the charge is completely delocalised; it might be drawn as:

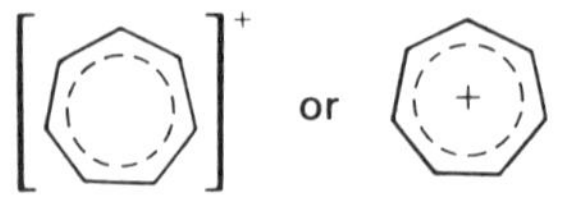

$$\left[\bigcirc\right]^+ \quad \text{or} \quad (+)$$

[We will meet this ion again in sections 10.6(c), 10.10(f).]

We have seen that a positive charge adjacent to an **unsaturated system** such as a double bond or a benzene ring is stabilised by delocalisation. The same is true of a positive charge adjacent to an atom (such as oxygen) with an **unshared pair of electrons**. Here are three examples:

(a) the ion produced by protonation of an aldehyde or ketone (and involved in some acid-catalysed reactions such as acetal formation)

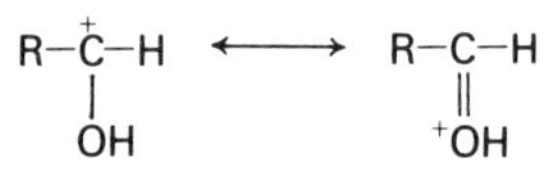

$$R-\overset{+}{C}H(OH) \longleftrightarrow R-CH{=}\overset{+}{O}H$$

A specific example of this type is the ion which we have already seen produced in the pinacol rearrangement [see section 1.7(b)]

(b) the ion derived from a chloromethyl ether:

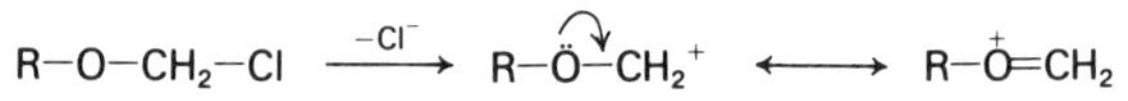

$$R-O-CH_2-Cl \xrightarrow{-Cl^-} R-\ddot{O}-CH_2^+ \longleftrightarrow R-\overset{+}{O}{=}CH_2$$

(c) the acylium ion involved in Friedel–Crafts acylation [see sections 4.4(e), 6.3(b)]

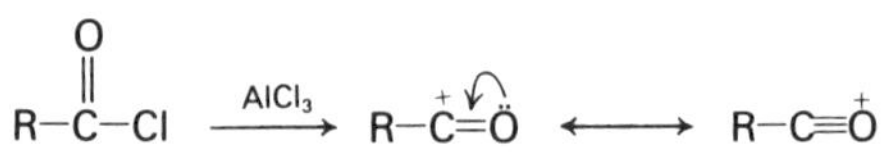

$$R-\overset{O}{\overset{\|}{C}}-Cl \xrightarrow{AlCl_3} R-\overset{+}{C}{=}\ddot{O} \longleftrightarrow R-C{\equiv}\overset{+}{O}$$

Note: In all these cases the resonance contribution drawn on the *right* is the more important of the two, as these all have atoms with a full share of electrons in their valence shells. Thus these ions would be better described as **oxonium** ions rather than carbocations.

1.10 Reactions involving allylic carbocations

As already mentioned allyl halides tend to dissociate relatively easily and thus can react by the S_N1 mechanism. For example, allyl bromide (3-bromopropene) gives a rapid precipitate with silver nitrate solution:

$$CH_2{=}CH-CH_2Br \rightleftharpoons CH_2{=}CH-CH_2^+ + Br^- \xrightarrow{Ag^+} AgBr$$

Another type of reaction which involves allylic carbocations is 1,4-addition to a conjugated diene.

e.g.

$$CH_2{=}CH{-}CH{=}CH_2 \xrightarrow{Br_2} \underset{\displaystyle Br}{CH_2}{-}\overset{+}{C}H{-}CH{=}CH_2 \longleftrightarrow \underset{\displaystyle Br}{CH_2}{-}CH{=}CH{-}CH_2^{+}$$

$$\downarrow Br^- \qquad\qquad \downarrow Br^-$$

$$\underset{\displaystyle Br}{CH_2}{-}\underset{\displaystyle Br}{CH}{-}CH{=}CH_2 \qquad \underset{\displaystyle Br}{CH_2}{-}CH{=}CH{-}\underset{\displaystyle Br}{CH_2}$$

(1,2-addition) (1,4-addition)

The initial attack of bromine to give a positively charged intermediate results in this case in an allylic carbocation. Thus it is easy to see how Br^- might subsequently add to either the 2- or 4-position of the ion resulting in the two observed products. (With excess bromine, of course, only one product is formed—1,2,3,4-tetrabromobutane.)

1.11 Reactions involving benzylic carbocations

Again because of the relative ease of dissociation benzyl halides (like allyl halides) can easily undergo reactions involving the S_N1 mechanism.

Here is another interesting reaction involving a benzylic carbocation. Imagine 1-phenyl-1-propene reacting with HBr under conditions which favour the ionic mechanism. [We shall consider what happens under conditions which promote the free radical mode of addition later in section 3.7.] There would be two possible products as shown here:

$$C_6H_5{-}CH{=}CH{-}CH_3 \xrightarrow{H^+} C_6H_5{-}\overset{+}{C}H{-}CH_2{-}CH_3 \xrightarrow{Br^-} C_6H_5{-}\overset{\displaystyle Br}{CH}{-}CH_2{-}CH_3$$

$$C_6H_5{-}CH{=}CH{-}CH_3 \xrightarrow{H^+} C_6H_5{-}CH_2{-}\overset{+}{C}H{-}CH_3 \xrightarrow{Br^-} C_6H_5{-}CH_2{-}\underset{\displaystyle Br}{CH}{-}CH_3$$

It is impossible to predict which would predominate simply from Markovnikov's rule, since both intermediate carbocations are secondary. In practice, however, the first product is formed exclusively, telling us that the first carbocation (a benzylic type) is the more stable of the two, a fact which we can rationalise as we have seen in terms of resonance delocalisation of the charge. (In the second carbocation, the positive charge is separated from the benzene ring by the intervening CH_2 group and cannot of course interact with it by resonance.)

2
Carbanions

2.1 Introduction

A carbanion is derived formally by the heterolytic cleavage of a C—H bond with the release of a proton.

e.g.

$$\mathrm{H}\overset{\cdot}{\underset{\cdot}{\div}}\overset{\displaystyle \mathrm{H}}{\underset{\displaystyle \mathrm{H}}{\mathrm{C}}}\overset{\cdot}{\underset{\cdot}{\div}}\mathrm{H} \longrightarrow \mathrm{H}\div\overset{\displaystyle \mathrm{H}}{\underset{\displaystyle \mathrm{H}}{\mathrm{C}}}\!:^{-} + \mathrm{H}^{+}$$

In practice, the tendency for this proton to dissociate from CH_4 (that is, its acidity) is negligible. The C—H bond is strong and (if it does break) the resulting negative charge is localised on the single carbon atom, not subject to any stabilising influences. This is reflected in the extremely high pK_a of methane ($\sim$42 or greater, difficult to estimate exactly). [For a definition of pK_a, see section 7.2.] Consequently a simple carbanion is highly reactive and is the strongest base we meet in organic chemistry; in fact we do not meet it very often. Most carbanions, as we shall see, are less reactive (i.e. more stable) because of resonance contributions which place the bulk of the negative charge on more electronegative elements such as oxygen, or because the species have considerable covalent character and are only 'carbanion-like' in their reactions (e.g. Grignard reagents).

2.2 Simple ionic carbanions

Only the most electropositive of metals can form ionic salts of alkanes such as R^- K^+. As mentioned above, the R^- in this salt is extremely basic and such a compound would be too indiscriminately reactive for much practical utility.

2.3 Alkynides

The non-acidity of an *alkene* is almost as impressive as that of an *alkane*, but it does become significantly easier to remove a proton as we progress to an *alkyne*, as Table 2.1 shows:

Table 2.1

	pK_a
ethane	42
ethene	36
ethyne	25

This is because the electrons in an *sp*-hybridised orbital are more closely held to the carbon nucleus than in an sp^2 or sp^3 orbital (that is, the *sp* orbital is more spherical in character). In other words *sp*-carbon is the most electronegative of the three types, since electronegativity is defined as the measure of an atom's ability to hold bonding electrons close to it. Consequently a proton is more easily lost from an *sp*-hybridised C—H bond than it is from an alkene or alkane.

Salts of terminal alkynes are formed by reaction with a strong base such as sodium amide in liquid ammonia.

e.g.

$$CH_3CH_2C{\equiv}CH + Na^+NH_2^- \rightarrow CH_3CH_2C{\equiv}C^-Na^+ + NH_3$$

The resulting alkynide ions are useful in organic syntheses as we shall see [sections 5.6(c), 6.4(b)].

2.4 Resonance-stabilised carbanions

We saw above [section 1.9] that a **positive** charge on a benzylic carbon atom is stabilised by delocalisation over the adjacent benzene ring. The same is true for a **negative** charge. Thus triphenylmethane (pK_a = 32) is rather more acidic than an ordinary alkane. The amide ion (NH_2^-) is just sufficiently basic to remove the proton from triphenylmethane:

$Ph_3C{-}H$	+	NH_2^-	$\rightleftharpoons$	Ph_3C^-	+	NH_3
(pK_a = 32)						(pK_a = 34)
(stronger acid)		(stronger base)		(weaker base)		(weaker acid)

When the proton is lost we can draw various resonance structures for the carbanion:

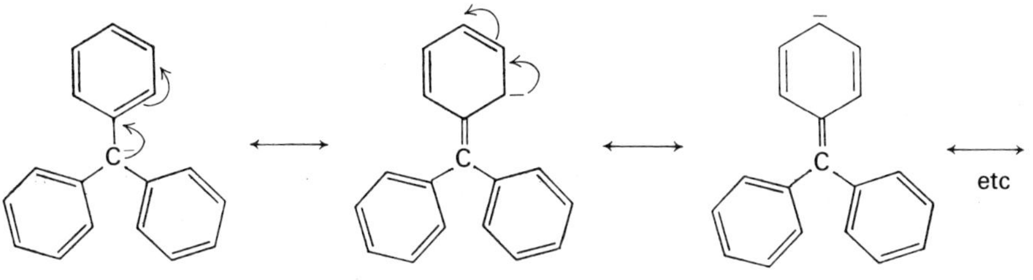

C ⟷ C ⟷ C ⟷ etc

Note that the carbanion is still a **very powerful base**. It can thus be used to remove a proton from hydrocarbons which are themselves more acidic than triphenylmethane.

$$R-C\equiv C-H + Ph_3C^- \rightarrow R-C\equiv C^- + Ph_3CH$$

$$C_5H_6 + Ph_3C^- \longrightarrow C_5H_5^- + Ph_3CH$$

The cyclopentadienyl anion in the second example can be represented by five exactly equivalent resonance structures (can you draw them?) and on this basis would be expected to be very stable. (As we will see later, the Molecular Orbital theory accounts for its unusual stability even better than does resonance theory [see section 10.6(b)].) In fact, 1,3-cyclopentadiene is one of the most acidic hydrocarbons known (pK_a = 15, more acidic than an alcohol).

2.5 Resonance-stabilised carbanions containing heteroatoms

So far we have looked at true carbanions, in the sense that the negative charge (either localised or delocalised) is carried only by **carbon**. Now we will extend our study to include those for which important resonance contributions can be written where the charge resides on a heteroatom such as O or N.

The most common sort of such carbanion arises upon the removal of a proton from a position alpha to a carbonyl group:

$$H-\overset{|}{\underset{|}{C}}-C(=O)- \xrightarrow{-H^+} >C^- -C(=O)- \longleftrightarrow >C=C(O^-)-$$

The resulting carbanion is stabilised by delocalisation of the negative charge on to the adjacent oxygen; in fact this resonance form is the more important of the two since it gives the charge to a very electronegative element (oxygen) better able to accommodate it than carbon. The species would be more correctly described as an oxyanion rather than a carbanion. Actually it is usually termed an **enolate** anion since it is the same species as formed from the corresponding enol form:

$$>C=C(OH)- \xrightarrow{-H^+} >C=C(O^-)- \longleftrightarrow >C^- -C(=O)-$$

Consider the pK_a values (of the underlined H atoms) in Table 2.2.

Thus a simple ketone [such as acetone, (propanone)] is more acidic than, say, ethyne (pK_a = 25) but a little less acidic than ethanol (pK_a = 18). We can also see that an ester group is less effective at stabilising the carbanion than is a keto

Table 2.2

	pK_a
$CH_3-C(=O)-OCH_2CH_3$	25
$CH_3-C(=O)-CH_3$	20
$CH_3CH_2O-C(=O)-CH_2-C(=O)-OCH_2CH_3$	13.3
$CH_3-C(=O)-CH_2-C(=O)-OCH_2CH_3$	10.7
$CH_3-C(=O)-CH_2-C(=O)-CH_3$	9.0

group (since an ester is not as acidic as a ketone) and we can explain this as follows. The oxygen atom in the carbonyl group of a ketone is electronegative—it tries to draw electrons towards itself—and it achieves a substantial fraction of a negative charge when it helps to stabilise the carbanion. The following is such an example.

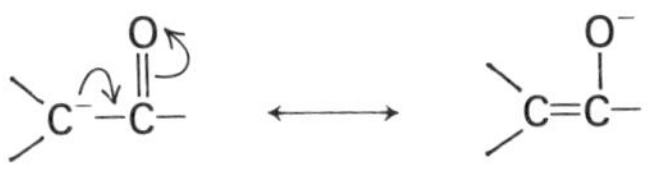

$$>\bar{C}-C(=O)- \longleftrightarrow >C=C(-O^-)-$$

However, in an ester, the oxygen of the carbonyl group *already* has its requirement for electrons partially fulfilled (by the other oxygen atom) as can be seen from this resonance description of its structure:

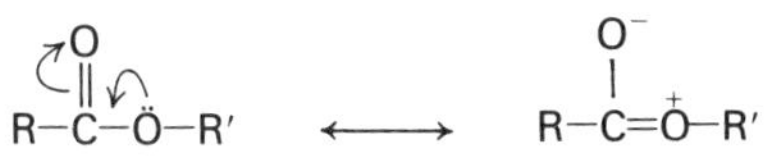

$$R-C(=O)-\ddot{O}-R' \longleftrightarrow R-C(-O^-)=\overset{+}{O}-R'$$

Therefore it has a smaller tendency to stabilise a negative charge at the α-position than does a ketone. [See also section 10.3(c).] It still does so to some extent of course; the pK_a of ethyl acetate (ethyl ethanoate) is much lower than that of an alkane, and the acidity of the α-hydrogens in an ester has important consequences in, for example, the Claisen condensation.

The other obvious conclusion from Table 2.2 is that a hydrogen atom alpha to *two* carbonyl groups is considerably more acidic than when it comes under the influence of only one. Here the possibilities of resonance–stabilisation are greater.

e.g.

$$CH_3-\overset{\overset{O}{\|}}{C}-\overset{-}{C}H-\overset{\overset{O}{\|}}{C}-CH_3 \longleftrightarrow CH_3-\overset{\overset{O^-}{|}}{C}=CH-\overset{\overset{O}{\|}}{C}-CH_3 \longleftrightarrow CH_3-\overset{\overset{O}{\|}}{C}-CH=\overset{\overset{O^-}{|}}{C}-CH_3$$

> **Connection** Both carbanions and carbocations can be stabilised by the presence of a heteroatom such as oxygen. In the case of a carbanion the O atom is *not directly* connected to the carbon carrying the negative charge but is one position removed:
>
> $$>\overset{-}{C}-\underset{|}{C}=O \longleftrightarrow >C=\underset{|}{C}-O^-$$
>
> In the case of a carbocation, and we saw examples in section 1.9, the O atom *is* directly connected to the carbon atom involved:
>
> $$>\overset{+}{C}-\ddot{O}- \longleftrightarrow >C=\overset{+}{O}-$$

Several other groups can increase the acidity of α-hydrogens (that is, stabilise the corresponding anions) in the same way as the carbonyl group. For instance the **nitro** group:

$$H-\overset{|}{\underset{|}{C}}-\overset{+}{N}(=O)O^- \xrightarrow{-H^+} >\overset{-}{C}-\overset{+}{N}(=O)O^- \longleftrightarrow >C=\overset{+}{N}(O^-)O^-$$

And the **cyano** group:

$$H-\overset{|}{\underset{|}{C}}-C\equiv N \xrightarrow{-H^+} >\overset{-}{C}-C\equiv N \longleftrightarrow >C=C=N^-$$

As expected, the more groups the more powerful is the effect:

Table 2.3

	pK_a	(Compare:)	pK_a
CH_3CN	25	H–C≡C–H	25
CH_3NO_2	10	Ph–OH	10
$CH_2(NO_2)_2$	3.6	CH_3COOH	4.8
$CH(CN)_3$	<1		

Thus dinitromethane, $CH_2(NO_2)_2$, is acidic enough to displace CO_2 from $NaHCO_3$ solution, and $CH(CN)_3$ has an acidity approaching that of the mineral acids.

Our last example of a group which can increase α-hydrogen acidity will be the **triphenylphosphonium** group:

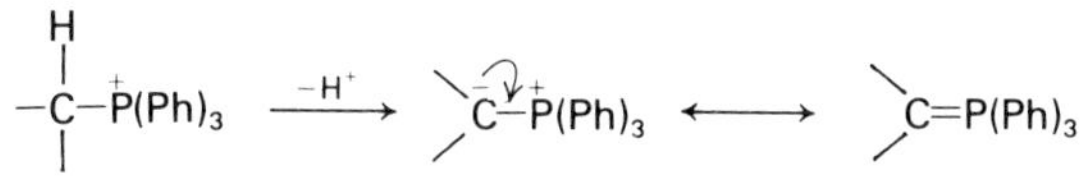

$$-\overset{|}{\underset{|}{C}}(H)-\overset{+}{P}(Ph)_3 \xrightarrow{-H^+} >\overset{-}{C}-\overset{+}{P}(Ph)_3 \longleftrightarrow >C=P(Ph)_3$$

(a quaternary phosphonium ion; compare: R_4N^+)

After the α-hydrogen is removed from the starting material (by a strong base) the product in this case is known as an **ylide**—this term is used for a species with a resonance contribution from a zwitterionic form as shown (i.e. a carbanion with a positive charge on an adjacent heteroatom).

To conclude this section let me stress the following 'connection'—whenever there is a hydrogen alpha to an electron-withdrawing group (capable of accepting electrons by resonance) then (i) that α-hydrogen is appreciably acidic, and (ii) the corresponding anion is relatively stable.

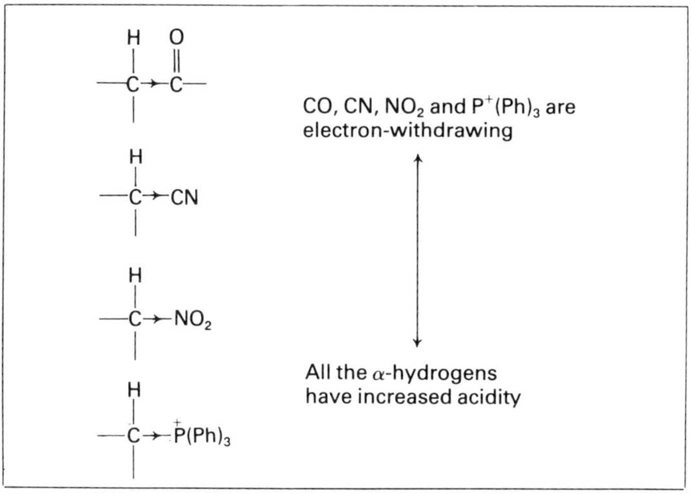

$$-\overset{H}{\underset{|}{C}}\rightarrow\overset{O}{\overset{\|}{C}}-$$

$$-\overset{H}{\underset{|}{C}}\rightarrow CN$$

$$-\overset{H}{\underset{|}{C}}\rightarrow NO_2$$

$$-\overset{H}{\underset{|}{C}}\rightarrow\overset{+}{P}(Ph)_3$$

CO, CN, NO_2 and $P^+(Ph)_3$ are electron-withdrawing

↕

All the α-hydrogens have increased acidity

Knowing this would enable you to predict the behaviour of compounds containing other electron-withdrawing groups, e.g.

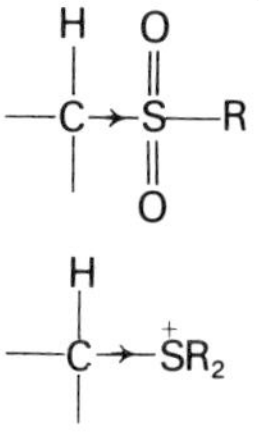

$$-\overset{H}{\underset{|}{C}}\rightarrow\overset{O}{\underset{O}{\overset{\|}{\underset{\|}{S}}}}-R$$

$$-\overset{H}{\underset{|}{C}}\rightarrow\overset{+}{S}R_2$$

2.6 Carbanions and keto–enol tautomerism

Consider the enolate anion formed when a base removes an α-hydrogen from, say, acetone (propanone):

$$CH_3-\overset{\overset{\displaystyle O}{\|}}{C}-CH_2^- \longleftrightarrow CH_3-\overset{\overset{\displaystyle O^-}{|}}{C}=CH_2$$

Now imagine a proton being replaced at either of the positions which carry the negative charge in the two hypothetical resonance forms. This gives us:

$$\underset{\text{(keto form)}}{CH_3-\overset{\overset{\displaystyle O}{\|}}{C}-CH_3} \quad \text{and} \quad \underset{\text{(enol form)}}{CH_3-\overset{\overset{\displaystyle OH}{|}}{C}=CH_2}$$

This is effectively what happens in practice. The equilibrium between keto and enol forms is rapidly set up through the intermediacy of the enolate anion when a trace of base is added.

Connection Keto–enol tautomerism and acidic α-hydrogens go hand in hand.

Here is another example:

Nitromethane

$\downarrow -H^+$

$$^-CH_2-\overset{+}{N}(=O)O^- \longleftrightarrow CH_2=\overset{+}{N}(O^-)O^-$$

$\downarrow H^+$

$$\underset{\text{nitro form}}{CH_3-\overset{+}{N}(=O)O^-} \quad \text{and} \quad \underset{\text{'aci' form}}{CH_2=\overset{+}{N}(OH)O^-}$$

(compare: a keto form and an enol form)

We can extend the connection. For a particular compound, the more resonance-stabilised the corresponding enolate ion is, the more acidic the α-hydrogen is, and the *greater the proportion of enol form at equilibrium*. Consider, for example, the data in Table 2.4.

Table 2.4

	pK_a	% enol
CH_3COCH_3	20	<<1
$CH_3COCH_2COCH_3$	9	~85

Clearly a *simple* ketone is much more stable than its enol form; the equilibrium lies overwhelmingly over to the keto side. This is not the same, however, for β-diketones. We have seen that the second carbonyl group gives added stability to the enolate ion (or in other words lowers the pK_a of the diketone) by resonance thus:

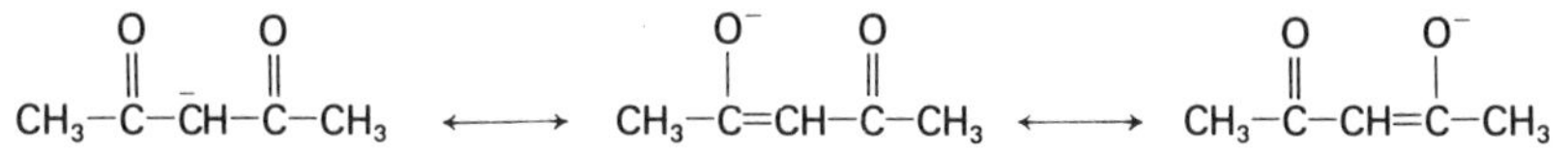

$$CH_3-\overset{\overset{O}{\|}}{C}-\overset{-}{C}H-\overset{\overset{O}{\|}}{C}-CH_3 \longleftrightarrow CH_3-\overset{\overset{O^-}{|}}{C}=CH-\overset{\overset{O}{\|}}{C}-CH_3 \longleftrightarrow CH_3-\overset{\overset{O}{\|}}{C}-CH=\overset{\overset{O^-}{|}}{C}-CH_3$$

In exactly the same way the enol form itself is resonance-stabilised by interaction with the second carbonyl group:

$$CH_3-\overset{\overset{:OH}{|}}{C}=CH-\overset{\overset{O}{\|}}{C}-CH_3 \longleftrightarrow CH_3-\overset{\overset{^+OH}{\|}}{C}-CH=\overset{\overset{O^-}{|}}{C}-CH_3$$

There is also a second reason for increased enol stability in cases such as this; it involves intramolecular hydrogen-bonding as in the following example.

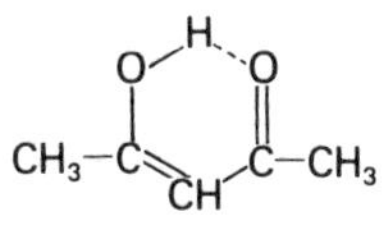

$$CH_3-\overset{\overset{O-H\cdots O}{|\qquad\quad\|}}{C=CH-C}-CH_3$$

Both these factors, resonance and hydrogen-bonding, can only apply to the enol form, and not to the keto form $CH_3COCH_2COCH_3$. Thus the greater proportion of enol in the β-diketone compared to the simple ketone is readily explicable.

2.7 Carbanions with some covalent character

As mentioned in section 2.1 very few simple carbanions could be considered to be totally ionic. Most metals form metal–carbon bonds with partial ionic and partial covalent character (although as we shall see, in writing reaction mechanisms we often consider these organometallic compounds as providers of carbanions).

(a) Organolithium compounds

e.g.

$$CH_3CH_2CH_2CH_2^-\ Li^+ \longleftrightarrow CH_3CH_2CH_2CH_2{-}Li$$

$$C_6H_5^-\ Li^+ \longleftrightarrow C_6H_5{-}Li$$

These reagents are rather like the more familiar Grignard reagents but are somewhat more reactive as they have a lot of 'carbanion character'.

(b) Grignard reagents

A Grignard reagent is usually written as RMgX where X is a halogen. In fact the structure is more complex than this, involving also R_2Mg and solvated species such as:

$$\begin{array}{c} R \\ | \\ Et_2O{-}{-}{-}{-}Mg{-}{-}{-}{-}OEt_2 \\ | \\ X \end{array}$$

Furthermore the R–Mg bond is polarised, which again we can represent by resonance.

$$CH_3{-}Mg^+\ Br^- \longleftrightarrow CH_3^-\ Mg^{++}\ Br^-$$

(c) Organocadmium reagents

We can describe the structure of organocadmium compounds as:

$$R{-}Cd{-}R \longleftrightarrow R^-\ {}^+Cd{-}R \longleftrightarrow R{-}Cd^+\ R^- \longleftrightarrow R^-\ Cd^{++}\ R^-$$

In this case the true nature lies rather closer to the covalent resonance contribution than it does with organomagnesium compounds. (In other words, organocadmium reagents are less 'carbanion-like' than Grignard reagents.) This is reflected in the following useful synthesis of ketones:

$$R'{-}COCl \xrightarrow{R_2Cd} R'{-}CO{-}R$$

The organocadmium is reactive enough to attack the acyl halide but not the resulting ketone. A Grignard reagent of course would go on to convert the ketone to a 3° alcohol [see sections 5.5(g), 6.5(a), (f)].

(Finally, less electropositive metals such as mercury form covalent metal–carbon bonds which have virtually no 'carbanion-like' contribution at all; e.g. $CH_3{-}Hg{-}CH_3$.)

2.8 Reactions of carbanions—Preview

In the following sections we will see carbanions (or carbanion-like species) acting in the following general ways:

(a) as bases:

$$R^- + H{-}A \rightarrow R{-}H + A^-$$

(b) as nucleophiles:
  (i) at a saturated carbon:

$$R^- + {>}C{-}X \rightarrow R{-}C{<} + X^-$$

  (ii) at a carbonyl group:

$$R^- + {>}C{=}O \rightarrow R{-}\overset{|}{\underset{|}{C}}{-}O^-$$

  (iii) at an α,β-unsaturated carbonyl group:

$$R^- + {>}C{=}\underset{|}{C}{-}\underset{|}{C}{=}O \rightarrow R{-}\overset{|}{\underset{|}{C}}{-}\underset{|}{C}{=}\underset{|}{C}{-}O^-$$

2.9 Reactions of carbanions as bases

Carbanions for the most part are very strong bases, or to put it another way they are the anions of very weak acids. [See also section 7.2.] Thus it is not surprising that one of their ready reactions is to combine with a proton whenever they can. Any compound which is a stronger acid than the parent hydrocarbon of the carbanion will supply the proton it needs. An example is the reaction of calcium carbide with water (used by cavers in their acetylene headlamps):

$$Ca^{++}\ {}^-C{\equiv}C^- + 2H_2O \rightarrow H{-}C{\equiv}C{-}H + Ca(OH)_2$$

$pK_a = 15.7$ (stronger acid) $\quad pK_a = 25$ (weaker acid)

Consider also what happens if moisture is allowed to get into a flask in which a Grignard reaction is to be carried out.

e.g.

$$CH_3^- \;(\text{from } CH_3MgBr) + H_2O \rightarrow CH_4 + OH^-$$

H_2O: $pK_a = 15.7$ $\quad CH_4$: $pK \sim 45$

The strong basicity of Grignard reagents must be borne in mind when planning reactions which involve compounds with acidic functional groups. For instance we all know that ketones react with Grignard reagents to give 3° alcohols, but consider ketones such as the following:

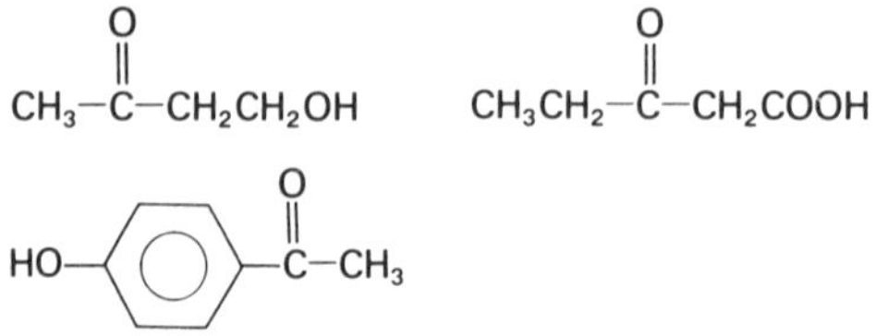

$$CH_3-\overset{\overset{O}{\|}}{C}-CH_2CH_2OH \qquad CH_3CH_2-\overset{\overset{O}{\|}}{C}-CH_2COOH$$

$$HO-C_6H_4-\overset{\overset{O}{\|}}{C}-CH_3$$

In these cases we first get acid–base reactions occurring between the carbanion-like Grignard reagent and the weakly or moderately acidic alcohol, carboxyl, and phenol groups.

e.g.

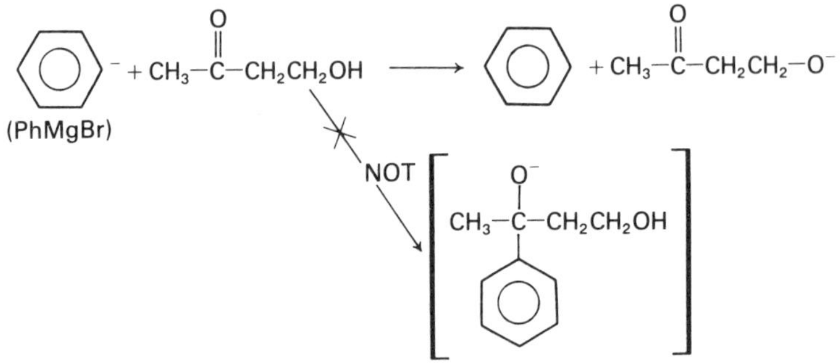

$$\underset{(PhMgBr)}{C_6H_5^-} + CH_3-\overset{\overset{O}{\|}}{C}-CH_2CH_2OH \longrightarrow C_6H_6 + CH_3-\overset{\overset{O}{\|}}{C}-CH_2CH_2-O^-$$

NOT

$$\left[CH_3-\underset{C_6H_5}{\overset{O^-}{\overset{|}{\underset{|}{C}}}}-CH_2CH_2OH\right]$$

(Attack at the carbonyl group may then follow if two equivalents of Grignard reagent are used.)

In fact, if you realise the 'connection' between R^- (of a Grignard reagent) and any acidic hydrogen atom, you might ask 'What about the acidic α-hydrogen of an aldehyde or ketone? Why does the Grignard reagent not react with that, ('that' refers to the α-hydrogen, not the aldehyde or ketone), rather than attack the carbonyl group nucleophilically?'

e.g.

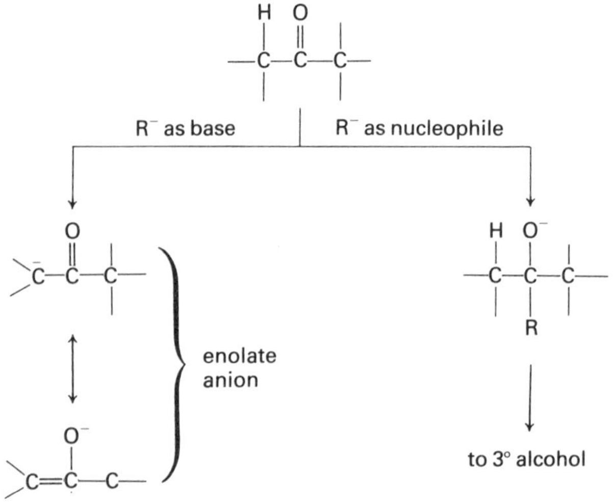

$$-\overset{H}{\overset{|}{\underset{|}{C}}}-\overset{O}{\overset{\|}{C}}-\overset{|}{\underset{|}{C}}-$$

R^- as base:

$$\left.\begin{array}{c} >\bar{C}-\overset{\overset{O}{\|}}{C}-\overset{|}{\underset{|}{C}}- \\ \updownarrow \\ >C=\overset{O^-}{\overset{|}{C}}-\overset{|}{C}- \end{array}\right\} \text{enolate anion}$$

R^- as nucleophile:

$$-\overset{H}{\overset{|}{\underset{|}{C}}}-\overset{O^-}{\overset{|}{\underset{R}{\underset{|}{C}}}}-\overset{|}{\underset{|}{C}}- \longrightarrow \text{to 3° alcohol}$$

The answer is that to some extent it *does*, but fortunately this is usually only a minor side-reaction which does not compete too drastically with the desired nucleophilic attack on the carbonyl group.

A similar situation arises in the reactions of sodium alkynides, $RC{\equiv}C^- Na^+$. The usual desired reaction is with $RC{\equiv}C^-$ acting as a nucleophile, and again the unwanted side-reaction is if it acts as a **base** instead.

e.g.

$$R{-}C{\equiv}C^- + H{-}\overset{|}{\underset{|}{C}}{-}\overset{|}{\underset{|}{C}}{-}Br \xrightarrow[\text{(substitution)}]{\text{as nucleophile}} R{-}C{\equiv}C{-}\overset{|}{\underset{|}{C}}{-}\overset{|}{\underset{|}{C}}{-}H + Br^-$$

$$R{-}C{\equiv}C^- + H{-}\overset{|}{\underset{|}{C}}{-}\overset{|}{\underset{|}{C}}{-}Br \xrightarrow[\text{(elimination)}]{\text{as base}} R{-}C{\equiv}C{-}H + \rangle C{=}C\langle + Br^-$$

If the alkyl halide is 1° then nucleophilic substitution (S_N2) occurs, but if it is 2° or 3° (and therefore sterically hindered) the strong alkynide base abstracts a proton and promotes elimination instead [section 7.7].

Of course, sometimes we *want* the carbanion to act as a base. We have already seen the example of sodium triphenylmethide acting as a base to remove a proton from 1,3-cyclopentadiene [section 2.4]. Here are some more examples of the use of carbanions (or organometallic compounds with carbanion-like character) acting as bases:

$$\underset{\substack{pK_a \sim 25 \\ \text{(stronger acid)}}}{CH_3(CH_2)_3C{\equiv}C{-}H} + \underset{(n\text{-butyllithium})}{CH_3CH_2CH_2CH_2Li} \rightarrow CH_3(CH_2)_3C{\equiv}C^-Li^+ + \underset{\substack{pK_a > 40 \\ \text{(weaker acid)}}}{CH_3CH_2CH_2CH_3}$$

$$\underset{pK_a \sim 40}{(CH_3CH_2CH_2)_2NH} + CH_3CH_2CH_2CH_2Li \rightarrow (CH_3CH_2CH_2)_2N^-Li^+ + \underset{pK_a > 40}{CH_3CH_2CH_2CH_3}$$

$$\underset{pK_a = 20}{CH_3{-}\overset{\overset{O}{\|}}{C}{-}CH_3} + Ph_3C^-K^+ \longrightarrow \left\{ CH_3{-}\overset{\overset{O^-}{|}}{C}{=}CH_2 \longleftrightarrow CH_3{-}\overset{\overset{O}{\|}}{C}{-}CH_2^- \right\} K^+ + \underset{pK_a = 32}{Ph_3C{-}H}$$

$$\mathrm{Me_2CH{-}\overset{\overset{\displaystyle O}{\|}}{C}{-}CMe_2{-}COOEt + Ph_3C^-Na^+ \longrightarrow \left.\begin{array}{c} Me_2C{=}\overset{\overset{\displaystyle O^-}{|}}{C}{-}CMe_2{-}COOEt \\ \updownarrow \\ Me_2\overset{-}{C}{-}\overset{\overset{\displaystyle O}{\|}}{C}{-}CMe_2{-}COOEt \end{array}\right\} Na^+ + Ph_3C{-}H}$$

[We will meet the last example again in section 7.6(c).]

2.10 Reactions of carbanions—Nucleophilic attack on alkyl halides

We have already seen that sodium alkynides can react with unhindered alkyl halides. The mechanism is S_N2 with $RC{\equiv}C^-$ acting as a nucleophile and the halide as the leaving group:

$$\mathrm{R{-}C{\equiv}C^- \curvearrowright R'{-}X \rightarrow R{-}C{\equiv}C{-}R' + X^-}$$

An *exactly analogous* reaction can be carried out with the resonance-stabilised carbanions we have examined in section 2.5 derived from β-dicarbonyl compounds such as ethyl acetoacetate (ethyl 3-oxobutanoate):

$$\mathrm{\begin{array}{c} CH_3CO \\ | \\ CH^- \\ | \\ COOEt \end{array} \curvearrowright R'{-}X \longrightarrow \begin{array}{c} CH_3CO \\ | \\ CH{-}R' \\ | \\ COOEt \end{array} + X^-}$$

The anion is derived by treatment of the starting material with a base such as sodium ethoxide. If desired the reaction can be repeated to introduce a second alkyl group:

$$\mathrm{\begin{array}{c} CH_3CO \\ | \\ R'{-}CH \\ | \\ COOEt \end{array} \xrightarrow{EtO^-} \begin{array}{c} CH_3CO \\ | \\ R'{-}C^- \\ | \\ COOEt \end{array} \curvearrowright R''{-}X \longrightarrow \begin{array}{c} CH_3CO \\ | \\ R'{-}C{-}R'' \\ | \\ COOEt \end{array} + X^-}$$

We will consider the synthetic utility of these reactions later in sections 6.4(d) and (e).

2.11 Reactions of carbanions—Nucleophilic attack on carbonyl compounds

This is an extremely important segment of organic chemistry. The carbonyl group is polar and the partial positive charge on the carbon atom provides an attractive target for all sorts of nucleophiles, including carbanions. We will see that for many reactions, the fundamental process underlying all of them can be represented as:

$$-\overset{|}{\underset{|}{C}}^{-} + {}^{\delta+}C=O^{\delta-} \longrightarrow -\overset{|}{\underset{|}{C}}-\overset{|}{\underset{|}{C}}-O^{-}$$

For instance, this is obviously true for the reactions of aldehydes and ketones with Grignard reagents:

$$\underset{(R-MgX)}{R^{-}} + C=O \longrightarrow R-\overset{|}{\underset{|}{C}}-O^{-}\ \overset{+}{M}gX \xrightarrow{H^{+}} R-\overset{|}{\underset{|}{C}}-OH$$

This constitutes an important and versatile method of synthesising 1°, 2° and 3° alcohols depending on whether we choose methanal, another aldehyde, or a ketone as the starting material. [See section 6.5(a).]

Carbon dioxide reacts with a Grignard reagent in the *same* basic way:

$$\underset{(R-MgX)}{R^{-}} + O=C=O \longrightarrow R-C(=O)O^{-}\ \overset{+}{M}gX \xrightarrow{H^{+}} R-COOH$$

An ester too reacts initially according to the *same* mechanism:

$$\underset{(RMgX)}{R^{-}} + \overset{OEt}{\underset{R'}{C}}=O \longrightarrow R-\overset{OEt}{\underset{R'}{C}}-O^{-}$$

This time, however, an alkoxide ion is then displaced and the resulting ketone then reacts further to give a 3° alcohol:

$$R-\overset{OEt}{\underset{R'}{C}}-O^{-} \longrightarrow R-\underset{R'}{C}=O + EtO^{-} \xrightarrow{RMgX} R-\overset{R}{\underset{R'}{C}}-OH$$

Other organometallic compounds can react similarly. We have already considered **organocadmium** reagents in the preparation of ketones [Section 2.7(c).]

e.g.

$$\underset{(Et_2Cd)}{CH_3CH_2^{-}} + C_6H_5-\overset{Cl}{C}=O \longrightarrow CH_3CH_2-\overset{Cl}{\underset{C_6H_5}{C}}-O^{-} \longrightarrow CH_3CH_2-\underset{C_6H_5}{C}=O + Cl^{-}$$

An **organozinc** reagent is involved in the **Reformatsky** reaction.
e.g.

$$\text{cyclohexanone} + BrCH_2COOEt \xrightarrow{Zn} \text{1-hydroxy-1-}(CH_2COOEt)\text{cyclohexane}$$

The intermediate organozinc compound may be considered to be:

$$Br^- Zn^{++}\ {}^-CH_2-\overset{O}{\overset{\|}{C}}-OEt \longleftrightarrow Br^- Zn^{++}\ CH_2=\overset{O^-}{\overset{|}{C}}-OEt \longleftrightarrow Br^-\overset{+}{Zn}-CH_2-\overset{O}{\overset{\|}{C}}-OEt$$

It is reactive enough to attack the polarised carbonyl group of a ketone but not the less polarised carbonyl of its own ester group. [See section 10.3(c).] Once again we are essentially looking at the *same* basic process:

$$EtO-CO-CH_2^- \curvearrowright \gt C=O \longrightarrow EtO-CO-CH_2-\overset{|}{\underset{|}{C}}-O^- \xrightarrow{H^+} EtO-CO-CH_2-\overset{|}{\underset{|}{C}}-OH$$

(from the organozinc compound)

Alkynides can attack aldehydes and ketones nucleophilically—the *same* type of reaction as all those in this section:

$$R-C\equiv C^- \curvearrowright \gt C=O \rightarrow R-C\equiv C-\overset{|}{\underset{|}{C}}-O^-$$

e.g.

$$HC\equiv C^-\ Na^+ + \text{cyclohexanone} \longrightarrow \text{1-}(HC\equiv C)\text{-1-}O^-\text{cyclohexane} \xrightarrow{H_2O} \text{1-}(HC\equiv C)\text{-1-}OH\text{cyclohexane}$$

Now we must consider all those **resonance-stabilised carbanions** (enolate anions) which can be derived by removal of an acidic α-hydrogen from a carbonyl compound:

$$-\overset{H}{\overset{|}{\underset{|}{C}}}-\overset{O}{\overset{\|}{C}}- \xrightarrow{-H^+} \gt\bar{C}-\overset{O}{\overset{\|}{C}}- \longleftrightarrow \gt C=\overset{O^-}{\overset{|}{C}}-$$

(A) (B)

Although resonance form B represents the real electronic structure of the ion more closely than A, in most of the subsequent reactions of an ion such as this it reacts as if it had structure A; [see section 10.3(c)]. Consider, for example, the **aldol condensation**, typified by the reaction of a simple aldehyde in the presence of dilute OH^-.

e.g.

$$2CH_3CH_2CHO \xrightarrow{OH^-} CH_3CH_2-C(H)(OH)-CH(CH_3)-CHO$$

The first step here is generation of the enolate ion:

$$CH_3CH_2C(=O)-H \underset{}{\overset{OH^-}{\rightleftharpoons}} CH_3-\bar{C}H-C(=O)-H \longleftrightarrow CH_3-CH=C(O^-)-H$$

This then attacks a second molecule of the aldehyde:

$$CH_3\bar{C}H(CHO) + H-C(=O)CH_2CH_3 \longrightarrow CH_3CH(CHO)-C(H)(O^-)CH_2CH_3 \xrightarrow{H^+} \text{product}$$

We have here just one more example of our key process:

$$-\overset{|}{\underset{|}{C}}^- \;\; >C=O \longrightarrow -\overset{|}{\underset{|}{C}}-\overset{|}{\underset{|}{C}}-O^-$$

Likewise in the **Claisen condensation**, the preparation of ethyl 3-oxobutanoate ('acetoacetic ester'), we have first the generation of an enolate anion:

$$CH_3-C(=O)-OEt \overset{EtO^-}{\rightleftharpoons} {}^-CH_2-C(=O)-OEt \longleftrightarrow CH_2=C(O^-)-OEt$$

Now here comes our key reaction *again*:

$$\bar{C}H_2(COOEt) + C(OEt)(=O)CH_3 \rightleftharpoons CH_2(COOEt)-C(OEt)(O^-)CH_3 \rightleftharpoons EtOOC-CH_2-C(=O)CH_3 + {}^-OEt$$

(i.e. CH_3COCH_2COOEt)

Similarly in the **Knoevenagel** reaction (the base-catalysed condensation of diethyl propanedioate, or diethyl malonate, with an aldehyde) we have first:

$$EtO-\overset{O}{\overset{\|}{C}}-CH_2-\overset{O}{\overset{\|}{C}}-OEt \xrightarrow[\text{(e.g. pyridine)}]{\text{base}} EtO-\overset{O}{\overset{\|}{C}}-\bar{C}H-\overset{O}{\overset{\|}{C}}-OEt \longleftrightarrow EtO-\overset{O}{\overset{\|}{C}}-CH=\overset{O^-}{\overset{|}{C}}-OEt \longleftrightarrow EtO-\overset{O^-}{\overset{|}{C}}=CH-\overset{O}{\overset{\|}{C}}-OEt$$

Then the resonance-stabilised carbanion attacks the carbonyl group of the aldehyde in once again the *same* mechanism as all those above:

$$(EtOOC)_2CH^- + \underset{Ph}{\overset{H}{}}C=O \longrightarrow (EtOOC)_2CH-\overset{H}{\underset{Ph}{C}}-O^- \xrightarrow{\text{protonation and dehydration}} (EtOOC)_2C=CH-Ph$$

Lastly, let us recall the resonance-stabilised carbanion known as a phosphorus ylide [see section 2.5].
e.g

$$^-CH_2-\overset{+}{P}(Ph)_3 \longleftrightarrow CH_2=P(Ph)_3$$

What then can this do to the carbonyl group of an aldehyde or ketone? Yes, of course, the phosphorus ylide can attack it nucleophilically in *just the same way* as all the previous examples:

$$\underset{^+P(Ph)_3}{\overset{|}{CH_2^-}} + \;C=O \longrightarrow \underset{^+P(Ph)_3}{CH_2}-\overset{|}{\underset{|}{C}}-O^-$$

In this particular case further reaction continues as follows:

$$\begin{array}{ll} CH_2 - C- \\ | \quad\quad | \\ (Ph)_3P^+ \;\; ^-O \end{array} \longrightarrow \begin{array}{ll} CH_2 - C- \\ | \quad\quad | \\ (Ph)_3P - O \end{array} \longrightarrow CH_2=C\langle \; + \; (Ph)_3P=O \; [\text{or } (Ph)_3\overset{+}{P}-O^-]$$

The products are triphenylphosphine oxide and an **alkene**—this process, known as the **Wittig** reaction, is an important mode of synthesis of alkenes.

e.g.

$$\text{cyclohexanone} + CH_2{=}P(Ph)_3 \longrightarrow \text{methylenecyclohexane}$$

$$(CH_3)_2C{=}O + CH_3{-}CH{=}P(Ph)_3 \longrightarrow (CH_3)_2C{=}CH{-}CH_3$$

Conclusion to this section

To learn about Grignard reactions, the aldol and Claisen and Knoevanagel condensations, the Reformatsky and Wittig reactions, and many others in isolation from one another would be an arduous and unfulfilling task. But spotting the common 'connection'—the nucleophilic attack of a carbanion R^- (or at least $R^{\delta-}$) on a carbonyl group—makes the learning process so much easier and more satisfying. I hope you agree.

2.12 Reactions of carbanions—Nucleophilic attack on α,β-unsaturated carbonyl compounds

An α,β-unsaturated carbonyl compound is usually written as:

$$>C{=}C{-}C{=}O$$

Other resonance contributors are:

$$>C{=}C{-}\overset{+}{C}{-}\overset{-}{O} \longleftrightarrow >\overset{+}{C}{-}C{=}C{-}\overset{-}{O}$$

This leads us to expect that a nucleophile such as a carbanion can attack not only at the carbonyl carbon (as dealt with in the preceding section) but also at the β-carbon (since this too is partially positively charged).

e.g.

$$\underset{(CH_3MgBr)}{CH_3^-} + CH_3CH{=}CH{-}\overset{O}{\overset{\|}{C}}{-}CH_3 \rightarrow CH_3CH{=}CH{-}C(OH)(CH_3){-}CH_3 \quad \text{simple addition} \quad (72\%)$$

$$+$$

$$CH_3CH(CH_3){-}CH_2{-}\overset{O}{\overset{\|}{C}}{-}CH_3 \quad \text{'conjugate' addition} \quad (20\%)$$

The following mechanism shows how the second product arises.

$$R^- + \overset{\delta+}{C}{=}C{-}C{=}\overset{\delta-}{O} \longrightarrow R{-}C{-}C{=}C{-}O^- \quad \text{(an enolate anion)}$$

$$\downarrow H_2O$$

$$R{-}C{-}C{=}C{-}OH \quad \text{(an enol)}$$

$$\downarrow$$

$$R{-}C{-}C(H){-}C{=}O \quad \text{(the keto form)}$$

Conjugate addition of **enolate anions** to α,β-unsaturated carbonyl compounds are known as **Michael** additions. In many cases conjugate addition predominates almost exclusively over simple addition to the carbonyl group.

e.g.

$$C_6H_5{-}CH{=}CH{-}COOEt + CH_2(COOEt)_2 \xrightarrow{EtO^-} C_6H_5{-}CH(CH(COOEt)_2){-}CH_2{-}COOEt$$

The base (ethoxide ion) removes an α-hydrogen from diethyl propanedioate ('malonic ester') to give the resonance-stabilised enolate ion which then attacks at the β-position of ethyl 3-phenylpropenoate (ethyl cinnamate). [For more examples, see section 5.8.]

3

Free Radicals

3.1 Introduction

We have started the previous two chapters by looking at the result of removing a hydride ion, H^-, or a proton, H^+, from a methane molecule to give a carbocation and a carbanion respectively. The third possibility is loss of a hydrogen atom, H·, and this gives us the methyl **free radical**:

$$H{:}\overset{H}{\underset{H}{\overset{\cdot\cdot}{\underset{\cdot\cdot}{C}}}}{:}H \rightarrow H{:}\overset{H}{\underset{H}{\overset{\cdot\cdot}{\underset{\cdot\cdot}{C}}}}\cdot + \cdot H$$

Once again (as with carbocations and carbanions) unless the free radical is subject to some significant stabilising influence, it is a very reactive short-lived species which only occurs fleetingly as an intermediate in some chemical reactions. Free radical reactions often occur in the gas phase, as under these circumstances **ionic** species are not solvated and are therefore of relatively high energy. On the other hand, ions in solution can be stabilised by interaction with solvent molecules and hence ionic reaction mechanisms are then frequently preferred to free radical ones.

3.2 Halogenation of methane

The first reaction in which students meet free radicals is often the reaction of methane with chlorine or bromine. (In fact, this is frequently the first reaction mechanism of any kind to be studied.) The mechanism involves several steps. First, there must be some kind of **initiation** process. This is the homolytic cleavage of a molecule to give free radicals or simply atoms such as Cl· atoms. **Homolysis** means the splitting of an electron-pair bond so that each fragment of the molecule retains one of the electrons (and **heterolysis** of course means that one fragment keeps both electrons and the other fragment gets neither).

In the chlorination of methane the initiation process is the homolysis of Cl_2 molecules to Cl atoms, brought about by heat or more usually by ultraviolet light:

$$:\ddot{\underset{\cdot\cdot}{Cl}}{:}\ddot{\underset{\cdot\cdot}{Cl}}: \xrightarrow{h\nu} :\ddot{\underset{\cdot\cdot}{Cl}}\cdot + \cdot\ddot{\underset{\cdot\cdot}{Cl}}:$$

(We usually write atoms or free radicals as X· concentrating on the unpaired electron and neglecting lone pairs.)

Next we get the **propagation** steps of the reaction occurring:

$$Cl\cdot + CH_4 \rightarrow CH_3\cdot + HCl$$
$$CH_3\cdot + Cl_2 \rightarrow CH_3{-}Cl + Cl$$

The Cl atom used up in the first of these processes is re-produced in the second and so this pair of reactions can continue almost indefinitely. This kind of situation (the repetition over and over again of a reaction or sequence of reactions) is known as a **chain reaction**. In this case the *net* process is:

$$CH_4 + Cl_2 \rightarrow CH_3{-}Cl + HCl$$

but remember, this equation does not represent the mechanism of the reaction, only its outcome.

I said above that the propagation steps go on almost indefinitely—almost, but not quite. After a single initiation reaction there may be hundreds or thousands of 'links in the chain' (i.e. propagation steps), but every so often we get **termination** steps occurring, where radicals or atoms are used up without fresh ones being re-formed. In this case, we may get as termination processes:

$$Cl\cdot + Cl\cdot \rightarrow Cl_2$$
$$Cl\cdot + CH_3\cdot \rightarrow CH_3Cl$$
$$CH_3\cdot + CH_3\cdot \rightarrow CH_3CH_3$$

A consideration of the energetics of the halogenation of methane is instructive. The rate-limiting step (i.e. the one which controls the overall rate of reaction) is the first propagation step:

$$X\cdot + CH_4 \rightarrow CH_3\cdot + H{-}X$$

Table 3.1 shows the enthalpy of reaction for this process for the four halogens.

Table 3.1

X	ΔH (kJ mol^{-1})
F	−134
Cl	+ 4.2
Br	+ 67
I	+138

For fluorine the reaction is very strongly exothermic with the result that the fluorination of methane is a dangerously vigorous or explosive process. For chlorine there is a low energy of activation for this step (~1 in 40 collisions between Cl· and CH_4 is productive at 275 °C) and the reaction proceeds smoothly. For bromine the activation energy barrier is higher—only ~1 in 10 million collisions at 275 °C is fruitful, but then there are numerous collisions happening, and the reaction still proceeds at an acceptable rate. For iodine, however, the process is so energetically unfavourable that a collision between I· and CH_4 virtually never has sufficient energy for reaction to occur. All that would happen in an attempted iodination of methane is initiation ($I_2 \rightarrow 2I\cdot$) followed by termination ($2I\cdot \rightarrow I_2$) without any intervening propagation.

3.3 Halogenation of other alkanes. Order of stability of free radicals

Everything in section 3.2 about the halogenation of methane applies more or less equally well to the halogenation of ethane, but when we move up to propane a new factor comes into play. The hydrogen atoms of propane are not all equivalent; six are primary and two are secondary. Thus if we were to consider the monohalogenation of propane, then from a simple statistical basis we would expect three times as much 1-substituted product as 2-substituted product. In practice, however, we get the following product distributions:

$$CH_3CH_2CH_3 \xrightarrow[h\nu,\ 25\,^\circ C]{Cl_2} \underset{45\%}{CH_3CH_2CH_2Cl} + \underset{55\%}{CH_3CHClCH_3}$$

$$CH_3CH_2CH_3 \xrightarrow[h\nu,\ 127\,^\circ C]{Br_2} \underset{3\%}{CH_3CH_2CH_2Br} + \underset{97\%}{CH_3CHBrCH_3}$$

Consider another example—2-methylpropane (isobutane). This has nine 1° hydrogens and one 3°, but we certainly do not get nine times as much 1° alkyl halide as 3°:

$$CH_3CH(CH_3)CH_3 \xrightarrow[h\nu,\ 25\,^\circ C]{Cl_2} \underset{64\%}{CH_3{-}CH(CH_3){-}CH_2Cl} + \underset{36\%}{CH_3{-}CCl(CH_3){-}CH_3}$$

$$CH_3CH(CH_3)CH_3 \xrightarrow[h\nu,\ 127\,^\circ C]{Br_2} \underset{<1\%}{CH_3{-}CH(CH_3){-}CH_2Br} + \underset{>99\%}{CH_3{-}CBr(CH_3){-}CH_3}$$

From these, and similar data, we can conclude that a 3° hydrogen is more easily abstracted in free radical substitution reactions than a 2° one which in turn is more easily abstracted than a 1° one. A corollary of this is that the resulting 3°

free radical is more stable than a 2° free radical which is more stable than a 1° free radical. Thus during the chlorination of propane, for example:

$$CH_3CH_2CH_3 \xrightarrow{Cl\cdot} H{-}Cl + CH_3CH_2CH_2\cdot$$

1° radical, less readily formed, less stable

$$CH_3CH_2CH_3 \xrightarrow{Cl\cdot} H{-}Cl + CH_3\dot{C}HCH_3$$

2° radical, more readily formed, more stable, preferred mode of reaction

Connection The order of stability of free radicals is the same as the order of stability of carbocations:

$$3° > 2° > 1°$$

The other obvious conclusion to be drawn from the product distributions above is that bromine atoms are more selective than chlorine atoms. That is, Br· seeks out the more easily abstracted H atoms more exclusively than Cl· does. We can explain this by considering the various transition states involved and making use of the postulate that the higher the energy of activation for the process, the later the transition state is reached in the progress of the reaction. Figure 3.1 will clarify this:

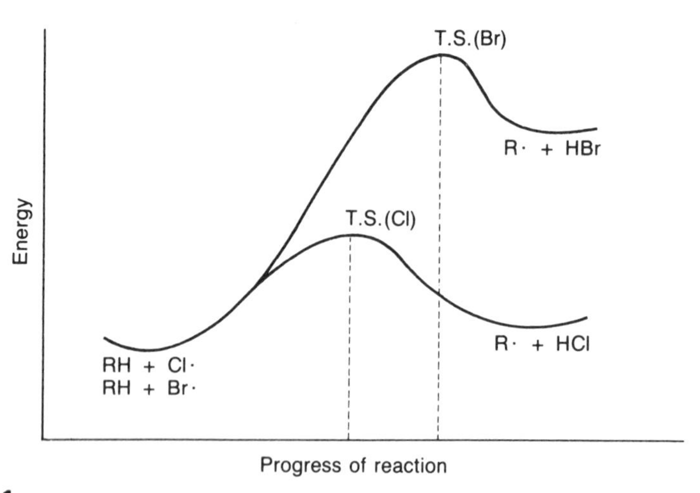

Fig. 3.1

Abstraction of a hydrogen atom by a bromine atom has a higher activation energy than abstraction by a chlorine atom [see section 3.2]; thus the transition state [T.S. (Br)] is reached later in the course of the transformation when bromine is involved than for chlorine [T.S. (Cl)]. This means that the **transition**

state more closely resembles the free radical product for this step of the bromination reaction than it does in chlorination. Accordingly, the relative energies of the transition states for abstraction of a 1°, 2°, or 3° hydrogen vary more widely for bromination than they do for chlorination (where the transition states have relatively less free radical character). The end result is that for both chlorination and bromination the ease of abstraction of H atoms follows the order 3° > 2° > 1°, but for bromination the trend is more dramatic (1600:80:1) than for chlorination (5:4:1).

Thus bromine atoms are not very reactive, but are highly selective, and chlorine atoms are more reactive, but less selective. Fluorine atoms are even more reactive, virtually every collision is fruitful and the relative ease of abstraction of 1°, 2°, and 3° hydrogen becomes insignificant (i.e. selectivity is zero). In fact, if fluorination of an alkane is carried out (very carefully!) the product distribution is approximately that expected on a purely statistical basis.

3.4 Addition of hydrogen bromide to alkenes

Recall the ionic mechanism for the addition of HBr to propene to give 2-bromopropane [see section 1.2(b)]:

$$CH_3CH{=}CH_2 \xrightarrow{HBr} CH_3\overset{+}{C}HCH_3 + Br^- \longrightarrow CH_3\underset{}{\overset{\overset{\displaystyle Br}{|}}{C}}HCH_3$$

This reaction proceeds as shown only in the absence of peroxides. In their presence, a different mechanism is involved and a different product is obtained (1-bromopropane). Historically this led to some confusion; sometimes the reaction would give one product, sometimes under seemingly identical conditions the other product would be obtained. We now know that organic chemicals (especially ethers) often contain traces of peroxides as impurities, and it is this factor which is responsible for the conflicting results.

How does a peroxide affect the outcome of this reaction? **By initiating a free radical mechanism**. The O—O bond in peroxides is quite weak and easily subject to homolysis by the influence of heat:

$$R{-}O{-}O{-}R \rightarrow 2R{-}O\cdot$$

The resulting radical then abstracts a hydrogen atom from H—Br:

$$RO\cdot + HBr \rightarrow ROH + Br\cdot$$

and next come the **propagation** steps:

$$Br\cdot + CH_3CH{=}CH_2 \rightarrow CH_3\dot{C}H{-}CH_2Br$$

$$CH_3\dot{C}H{-}CH_2Br + HBr \rightarrow CH_3CH_2CH_2Br + Br\cdot$$

Occasionally there is a **termination** reaction such as:

$$CH_3\dot{C}HCH_2Br + Br\cdot \rightarrow CH_3CHBrCH_2Br$$

The Br· atom used up in the first propagation step is re-formed in the second propagation step. This is therefore another example of a chain reaction and it is self perpetuating once it has been initiated by a trace of peroxide.

Connection The addition of HBr to alkenes in the presence of peroxides and the halogenation of alkanes are both free radical chain reactions.

Consider again the first propagation step. We could represent it in more detail as:

$$CH_3CH{=}CH_2 \; \cdot Br \rightarrow CH_3\dot{C}H{-}CH_2{-}Br$$

With free radicals we use a **half-headed arrow** to symbolise the movement of **one** electron. Contrast this with the first step in the alternative **ionic** mechanism where we use an ordinary arrow to mean the shifting of an electron-**pair**:

$$CH_3{-}CH{=}CH_2 \; H{-}Br \rightarrow CH_3{-}\overset{+}{C}H{-}CH_3 + Br^-$$

There remains the question of why the free radical process results in 1-bromopropane. Conceivably in the first propagation step the Br· atom could add to either end of the double bond:

$$CH_3CH{=}CH_2 \xrightarrow{Br\cdot} CH_3\dot{C}H{-}CH_2Br$$

or

$$CH_3CH{=}CH_2 \xrightarrow{Br\cdot} CH_3CH(Br){-}CH_2\cdot$$

The upper radical is 2°, the lower one is 1°. The fact that we get the product (1-bromopropane) derived from the 2° radical tells us that the 2° radical is more stable than the 1°. This is *exactly the same conclusion* we have drawn from looking at the products from the halogenation of propane [section 3.3]—another good example of a 'connection' between two different reactions because they both involve the same kind of mechanism and intermediates.

Superficially, you might think that the two modes of addition of HBr to propene are completely different—the first has an ionic mechanism, the second involves free radicals; the first gives 2-bromopropane, the second 1-bromopropane; the first follows Markovnikov's rule, the second gives the anti-Markovnikov product. However, there is one important common factor connecting the two processes—we can explain the outcome in both cases in terms of the **relative stabilities of the intermediates involved**. In the first case the

2° carbocation (formed from H^+ and propene) is more stable than the 1°, and in the second case the 2° free radical (formed from Br· and propene) is again more stable than the 1°.

3.5 Allyl free radicals

How does bromine react with propene? Your answer, I am sure, is that it **adds**:

$$CH_3CH{=}CH_2 + Br_2 \rightarrow CH_3\underset{|}{\overset{Br}{C}}H{-}\overset{Br}{\underset{|}{C}}H_2$$

And so, of course, it does—at least if propene is bubbled into a solution of bromine in for example CCl_4. However, under some reaction conditions a substitution happens instead:

$$CH_3CH{=}CH_2 + Br_2 \xrightarrow[\text{high temperature}]{\text{gas phase}} Br{-}CH_2CH{=}CH_2 + HBr$$

We can rationalise the change from one reaction to the other by considering their mechanisms. The first involves ionic intermediates [section 4.3(f)] and so is disfavoured in the gas phase (no solvation can occur). The second involves a **free radical mechanism** which is not disfavoured in the gas phase; moreover we are dealing here with a **resonance-stabilised** free radical.

The same kinds of processes (**initiation**, **propagation**, **termination**) happen in the gas-phase halogenation of propene as we saw happen in the halogenation of alkanes. In particular, the propagation steps are:

$$CH_3{-}CH{=}CH_2 + Br\cdot \rightarrow H{-}Br + \cdot CH_2{-}CH{=}CH_2$$

$$\cdot CH_2{-}CH{=}CH_2 + Br_2 \rightarrow Br{-}CH_2{-}CH{=}CH_2 + Br\cdot$$

The free radical involved here is an allyl radical—this is one where the odd electron is adjacent to a double bond. In fact, this electron is not localised on a single carbon atom as drawn above, but rather *delocalised* by interaction with the double bond:

$$\cdot CH_2{-}CH{=}CH_2 \longleftrightarrow CH_2{=}CH{-}\dot{C}H_2$$

This makes the allyl free radical more stable than even a tertiary radical and consequently reactions involving allyl radicals go relatively easily.

Connection Does the allyl free radical remind you of anything? It should—we have seen that the allyl **carbocation** is resonance-stabilised in just the same way [section 1.9]:

$$\overset{+}{C}H_2{-}CH{=}CH_2 \longleftrightarrow CH_2{=}CH{-}\overset{+}{C}H_2$$

3.6 Benzyl free radicals

Just as allyl radicals and allyl carbocations are 'connected' by both being resonance-stabilised, so also are benzyl radicals and carbocations [see section 1.9]. The unpaired electron on a benzyl radical can be delocalised round the benzene ring:

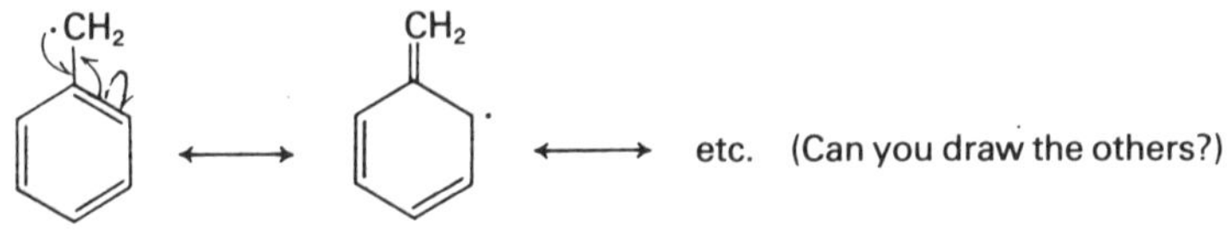

$\cdot CH_2$ ⟷ CH_2 ⟷ etc. (Can you draw the others?)

The extra stability of benzylic radicals accounts for (or more correctly, is suggested by) certain observations such as:

(a) the methyl side chain of toluene is halogenated under milder conditions than methane itself
(b) free radical halogenation of ethylbenzene gives solely 1-bromo-1-phenylethane (not 2-bromo-1-phenylethane):

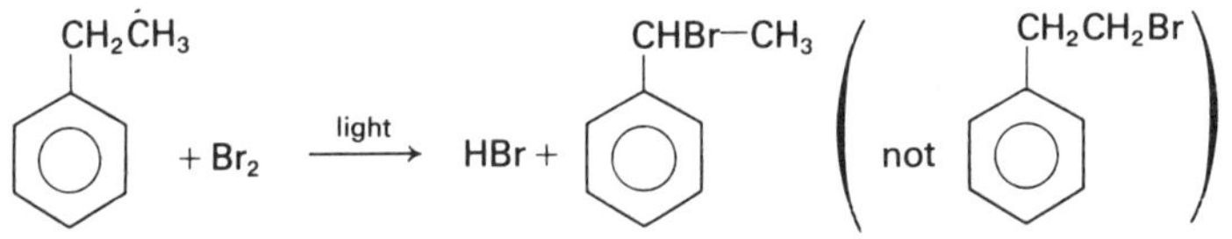

$$C_6H_5CH_2CH_3 + Br_2 \xrightarrow{\text{light}} HBr + C_6H_5CHBr{-}CH_3 \quad (\text{not } C_6H_5CH_2CH_2Br)$$

In this example, the intermediate free radical must be:

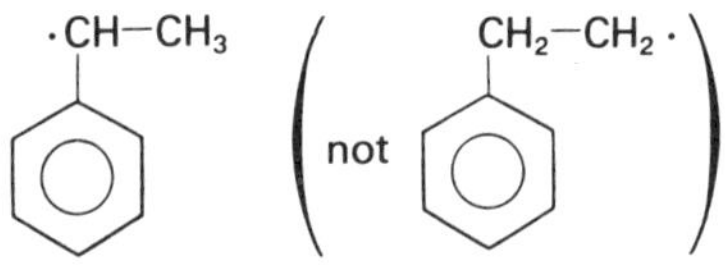

$$C_6H_5\dot{C}H{-}CH_3 \quad (\text{not } C_6H_5CH_2{-}CH_2\cdot)$$

This preferred free radical is not only 2°, as opposed to 1°, but is specifically a benzylic radical and hence resonance-stabilised. (Chlorination of ethylbenzene also gives mainly the 2° isomer, but a little of the 1° isomer is formed as well. Chlorine atoms are energetic enough occasionally to abstract the less reactive hydrogen at the non-benzylic position.)

3.7 Addition of HBr to Ph.CH=CH.CH$_3$

We have covered this reaction when it follows an *ionic* mechanism previously [section 1.11]. Under conditions which favour free radicals, however, the mechanism is:

Initiation

Rad · (e.g. from a peroxide) + HBr → Rad—H + Br ·

Propagation

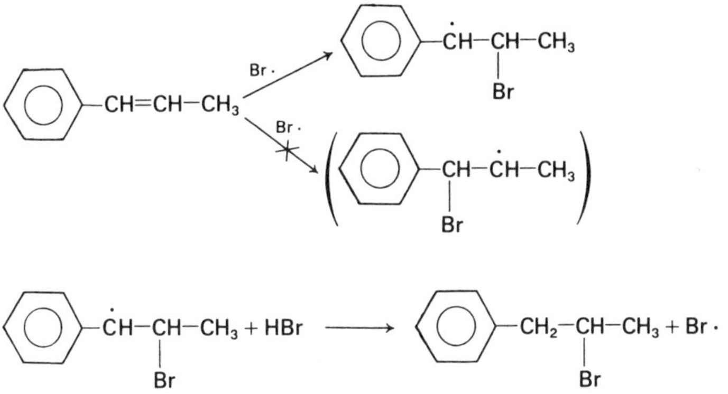

$$PhCH{=}CH{-}CH_3 \xrightarrow{Br\cdot} Ph\dot{C}H{-}CH(Br){-}CH_3 \quad \left(\not\rightarrow PhCH(Br){-}\dot{C}H{-}CH_3\right)$$

$$Ph\dot{C}H{-}CH(Br){-}CH_3 + HBr \longrightarrow PhCH_2{-}CH(Br){-}CH_3 + Br\cdot$$

In practice there is only one product found (1-phenyl-2-bromopropane). This tells us that the upper of the two radicals drawn above for the first propagation step is preferred (and the one in parentheses is not formed). This is because the preferred one is a **benzylic** radical; the odd electron is delocalised over the adjacent benzene ring.

Perhaps it will help to see how the benzylic free radical arises if we use half-headed arrows as follows.

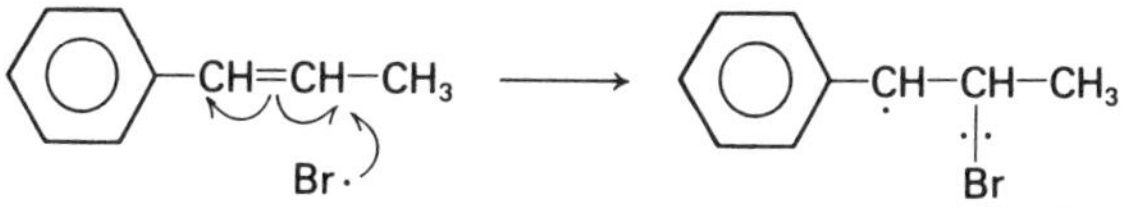

$$PhCH{=}CH{-}CH_3 + Br\cdot \longrightarrow Ph\dot{C}H{-}CH(Br){-}CH_3$$

(When Br· adds to one carbon of the double bond, a lone electron ends up on the other carbon.)

Connection The two apparently different reactions:

$$Ph{-}CH_2CH_3 \xrightarrow[\text{light}]{Br_2} Ph{-}CHBrCH_3 \quad (\text{not } PhCH_2CH_2Br)$$

$$PhCH{=}CHCH_3 \xrightarrow[\text{peroxides}]{HBr} PhCH_2CHBrCH_3 \quad (\text{not } PhCHBrCH_2CH_3)$$

are both trying to tell us *exactly the same thing*—that a benzylic free radical is unusually stable.

3.8 The triphenylmethyl free radical

Just as the triphenylmethyl carbocation is substantially more resonance-stabilised than the simple benzyl carbocation [sections 1.9, 1.10(d)], so is the triphenylmethyl radical much more stable than the benzyl radical:

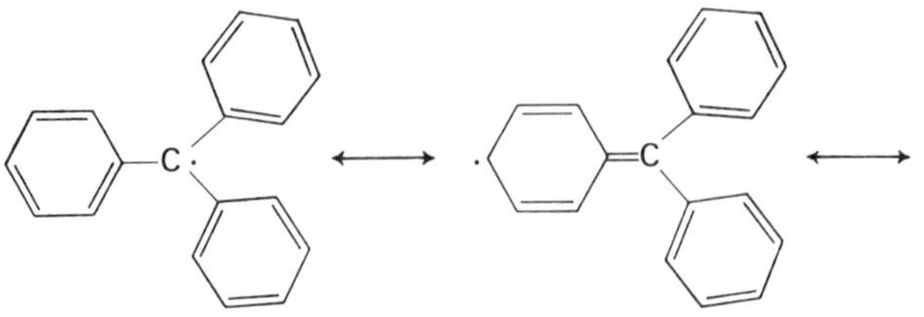

etc. (Do check you can draw them all.)

This species, it should be stressed, is *untypically* stable and long-lived compared to simple free radicals. In fact, the dimer of this radical (which was long thought to be hexaphenylethane, $Ph_3C{-}CPh_3$) dissociates quite markedly at room temperature whereas, of course, ethane does not.

$$\text{Dimer} \rightleftharpoons 2\,Ph_3C\cdot \quad \Delta H = 46\ \text{kJ mol}^{-1}$$

$$CH_3CH_3 \not\rightleftharpoons 2\,CH_3\cdot \quad \Delta H = 368\ \text{kJ mol}^{-1}$$

There are two reasons why the dimer dissociates so readily. Firstly, it is to relieve the steric crowding of six bulky phenyl groups in a single molecule, and secondly, because of the pronounced resonance–stabilisation of the $Ph_3C\cdot$ radical we have just been considering. (Neither of these factors applies to ethane, and its C—C bond is much stronger.)

In 1968 the dimer was shown not to be hexaphenylethane at all, but rather:

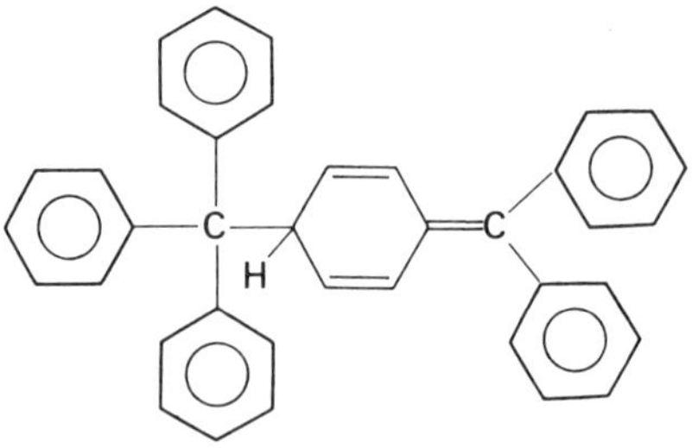

which can be thought of as arising from the coupling together of the two resonance forms of the triphenylmethyl radical which are drawn above, remembering of course that these two resonance forms do not represent different actual species [see section 10.1].

3.9 Oxidation of alkylbenzenes

Consider the following reactions:

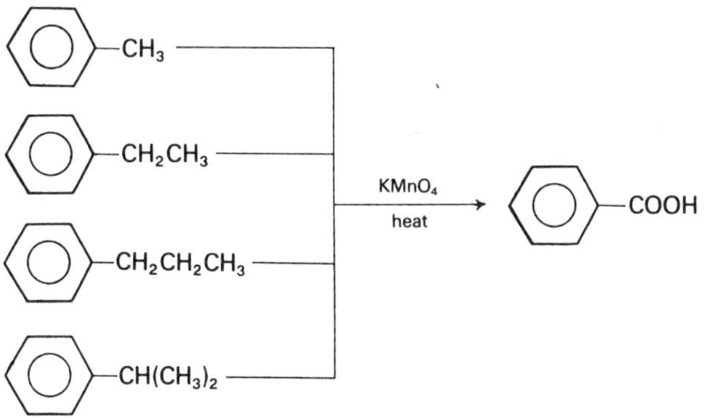

$C_6H_5-CH_3$, $C_6H_5-CH_2CH_3$, $C_6H_5-CH_2CH_2CH_3$, $C_6H_5-CH(CH_3)_2$ $\xrightarrow[\text{heat}]{KMnO_4}$ C_6H_5-COOH

and the following *non*-reactions:

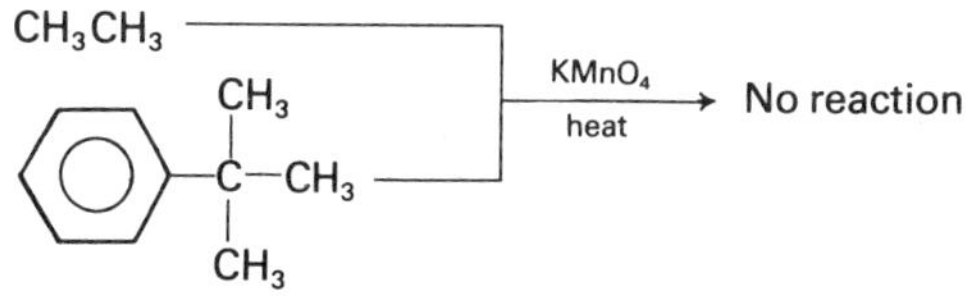

CH_3CH_3, $C_6H_5-C(CH_3)_3$ $\xrightarrow[\text{heat}]{KMnO_4}$ No reaction

From results such as these we can conclude that **side-chain oxidations are similar to benzylic halogenations** because in both cases the first step is abstraction of a benzylic hydrogen. If there is a benzylic hydrogen, then (with $KMnO_4$) the carbon at this position ends up as —COOH, and any other carbons are removed from the molecule. If there is no benzylic hydrogen (e.g. *t*-butylbenzene, or 2-methyl-2-phenylpropane) then there is no reaction. A simple alkane (e.g. ethane) does not have a reactive benzylic position of course, and so in this case there is also no reaction.

Connection

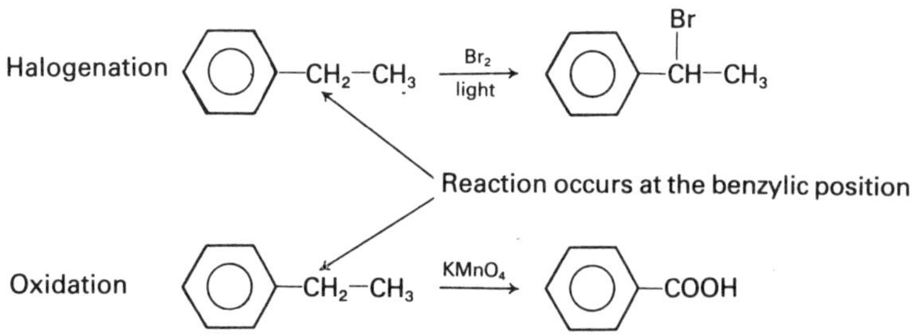

Halogenation $C_6H_5-CH_2-CH_3 \xrightarrow[\text{light}]{Br_2} C_6H_5-CH(Br)-CH_3$

Reaction occurs at the benzylic position

Oxidation $C_6H_5-CH_2-CH_3 \xrightarrow{KMnO_4} C_6H_5-COOH$

If there is more than one alkyl side chain then we get more than one —COOH group produced, e.g.

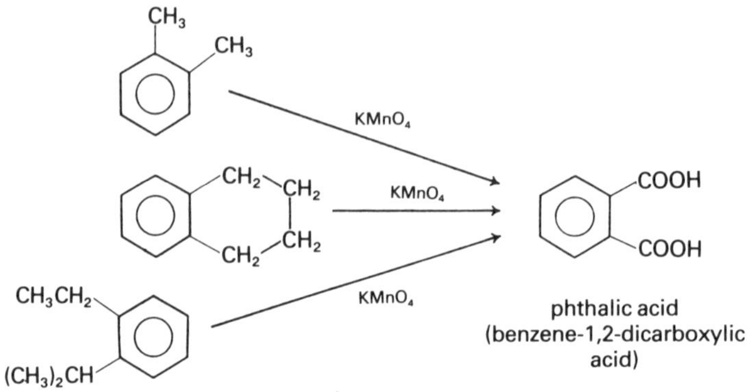

$$\text{o-xylene} \xrightarrow{KMnO_4} \text{phthalic acid}$$

$$\text{tetralin} \xrightarrow{KMnO_4} \text{phthalic acid}$$

$$\text{1-ethyl-2-isopropylbenzene} \xrightarrow{KMnO_4} \text{phthalic acid}$$

phthalic acid
(benzene-1,2-dicarboxylic acid)

3.10 Free radical polymerisation of alkenes

We have looked at the peroxide-promoted free radical mechanism for the addition of HBr to an alkene [section 3.4]. The first steps are:

$$\text{Peroxide} \rightarrow 2\,\text{Rad}\cdot$$
$$\text{Rad}\cdot + HBr \rightarrow \text{Rad}{-}H + Br\cdot$$

But what happens if there is no HBr present? Now the radical from the peroxide can **add to the alkene** instead of abstracting a hydrogen atom from HBr:

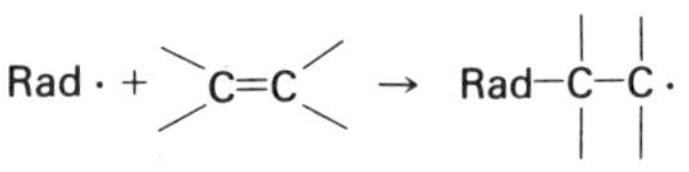

$$\text{Rad}\cdot + {>}C{=}C{<} \rightarrow \text{Rad}{-}\overset{|}{\underset{|}{C}}{-}\overset{|}{\underset{|}{C}}\cdot$$

Now we have a different radical which can add to another molecule of alkene:

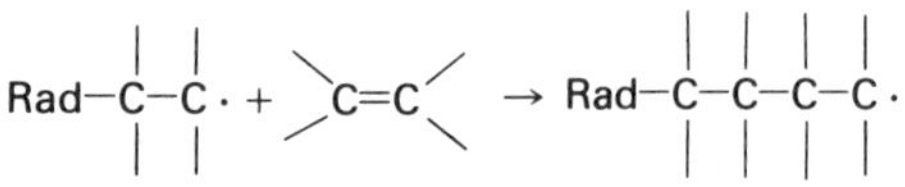

$$\text{Rad}{-}\overset{|}{\underset{|}{C}}{-}\overset{|}{\underset{|}{C}}\cdot + {>}C{=}C{<} \rightarrow \text{Rad}{-}\overset{|}{\underset{|}{C}}{-}\overset{|}{\underset{|}{C}}{-}\overset{|}{\underset{|}{C}}{-}\overset{|}{\underset{|}{C}}\cdot$$

and then this adds to another, and so on, for numerous repeats, until we have a very long polymeric chain. If the starting alkene is ethene then the product is variously known as polyethene, polyethylene or polythene.

Other alkenes can be similarly polymerised to give a multitude of useful materials with varying properties.

e.g.

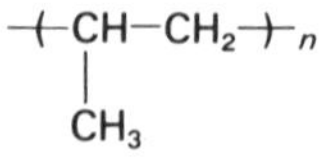

$$\left(\underset{\underset{CH_3}{|}}{CH}{-}CH_2 \right)_n \quad \text{polypropene, polypropylene}$$

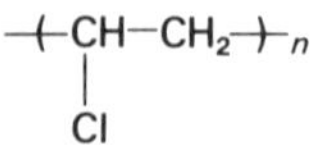

$$\left(\underset{\underset{Cl}{|}}{CH}{-}CH_2 \right)_n \quad \text{polyvinyl chloride, PVC}$$

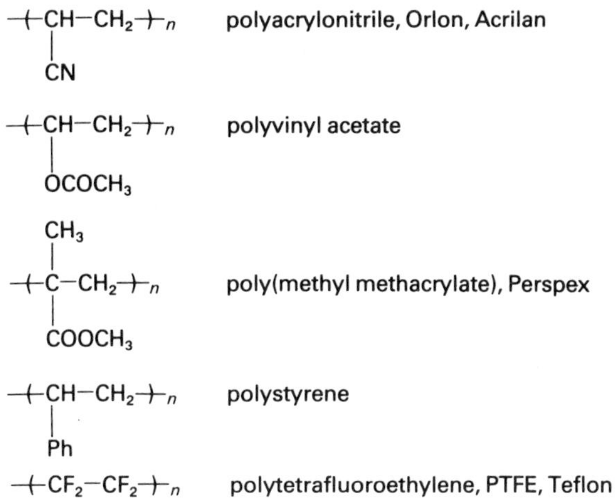

$-\!\!\!+\!CH(CN)-CH_2\!+\!\!\!-_n$ polyacrylonitrile, Orlon, Acrilan

$-\!\!\!+\!CH(OCOCH_3)-CH_2\!+\!\!\!-_n$ polyvinyl acetate

$-\!\!\!+\!C(CH_3)(COOCH_3)-CH_2\!+\!\!\!-_n$ poly(methyl methacrylate), Perspex

$-\!\!\!+\!CH(Ph)-CH_2\!+\!\!\!-_n$ polystyrene

$-\!\!\!+\!CF_2-CF_2\!+\!\!\!-_n$ polytetrafluoroethylene, PTFE, Teflon

3.11 Carboxylate radicals

We have noted that peroxides readily dissociate homolytically:

$$R-O-O-R \rightarrow 2R-O\cdot$$

A common example of this is with a diacyl peroxide such as benzoyl peroxide:

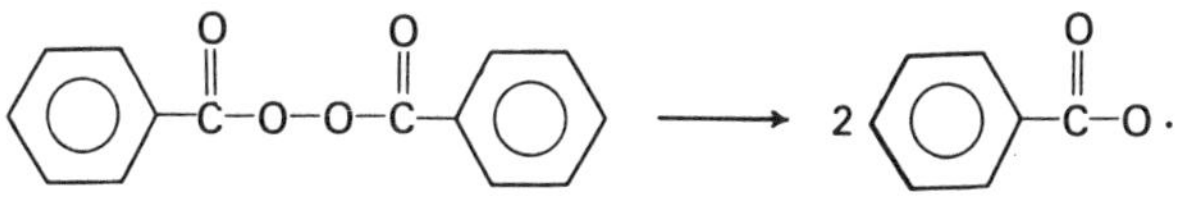

$$C_6H_5-\overset{O}{\overset{\|}{C}}-O-O-\overset{O}{\overset{\|}{C}}-C_6H_5 \longrightarrow 2\,C_6H_5-\overset{O}{\overset{\|}{C}}-O\cdot$$

The resulting carboxylate radical in turn readily dissociates to carbon dioxide and a phenyl radical:

$$C_6H_5-\overset{O}{\overset{\|}{C}}-O\cdot \longrightarrow C_6H_5\cdot + CO_2$$

This is why benzoyl peroxide is often used to initiate the free radical addition of HBr to alkenes, or the free radical polymerisation of alkenes such as styrene (phenylethene).

Another route to a carboxylate radical is by electrolysis of an aqueous solution of a sodium salt of a carboxylic acid. This reaction occurs at the anode:

$$R-\overset{O}{\overset{\|}{C}}-O^- \rightarrow R-\overset{O}{\overset{\|}{C}}-O\cdot + e^-$$

Again the carboxylate radicals give off CO_2 and the resulting R· radicals find themselves adsorbed in close proximity to each other on the surface of the electrode—hence they tend to combine before any other reaction can happen:

$$R\cdot + R\cdot \rightarrow R{-}R$$

This constitutes Kolbé's electrolytic alkane synthesis.

If, therefore, we write two such apparently different reactions as:

$$2CH_3CH_2CH_2CH_2COO^- \xrightarrow{\text{electrolysis}} CH_3(CH_2)_6CH_3$$

and

$$\text{styrene} \xrightarrow{\text{benzoyl peroxide}} \text{polystyrene}$$

it is satisfying to recognise that they are 'connected'.

3.12 Résumé of reactions of free radicals

A free radical can

(1) combine with another free radical
(2) abstract an atom from a molecule (to give a different molecule and a different free radical)
(3) add to an unsaturated molecule (to give a different free radical)
(4) decompose to two fragments (one of which is itself a free radical)

Reaction (1) is usually statistically unlikely as the concentration of free radicals is normally very low. Some examples are certain termination processes we have seen:

$$Br\cdot + Br\cdot \rightarrow Br_2$$

$$CH_3\cdot + CH_3\cdot \rightarrow CH_3CH_3$$

and in the Kolbé electrolytic reaction.

Examples of (2) are:

$$RO\cdot + H{-}Br \rightarrow ROH + Br\cdot$$

$$Br\cdot + R{-}H \rightarrow H{-}Br + R\cdot$$

$$R\cdot + Cl_2 \rightarrow R{-}Cl + Cl\cdot$$

Examples of (3) are:

$$Br\cdot + {>}C{=}C{<} \rightarrow Br{-}\overset{|}{\underset{|}{C}}{-}\overset{|}{\underset{|}{C}}\cdot$$

$$R\cdot + {>}C{=}C{<} \rightarrow R{-}\overset{|}{\underset{|}{C}}{-}\overset{|}{\underset{|}{C}}\cdot$$

An example of (4) is:

$$R{-}COO\cdot \rightarrow R\cdot + CO_2$$

You should now glance back through the preceding sections to see where these various examples occur.

4

Electrophilic Attack

4.1 Introduction

There is no such thing as electrophilic attack! By this provocative statement, I mean there is no such thing as electrophilic attack *on its own* (and equally, nucleophilic attack also cannot happen by itself). Consider this general reaction:

$$\underset{\text{(can furnish an electron pair)}}{N} + \underset{\text{(can accept an electron pair)}}{E} \rightarrow N{-}E$$

By definition N is a nucleophile (a reagent which seeks out positive charges) and E is an electrophile (a reagent looking for electrons). Therefore this reaction is not solely nucleophilic attack of N on E, or electrophilic attack of E on N; it is both. (Just as a 100 cm^3 flask containing 50 cm^3 of water is not just either half full or half empty—it is both.)

Why then do we often classify reactions as being electrophilic or nucleophilic? Purely as a matter of habit or convention. For example, if in the general equation above, N is organic and E is an **in**organic reagent then as narrow minded organic chemists we would tend to speak of this as **electrophilic** attack *of* E *on* our organic compound. The nitration of benzene illustrates this:

$$\underset{\text{(electron rich)}}{C_6H_6} \quad + \quad \underset{\text{(electron deficient)}}{NO_2^+}$$

We would usually say that the nitronium ion attacks benzene **electrophilically**. (But we would be perfectly justified in describing the reaction as **nucleophilic** attack *of* benzene *on* NO_2^+ instead.)

On the other hand, if N is inorganic and E organic then we generally speak of **nucleophilic** attack of N on our organic compound.

e.g.

$$\ddot{N}H_3 + CH_3{-}\overset{\overset{O^{\delta-}}{\|}}{\underset{\delta+}{C}}{-}Cl$$

The ammonia attacks the carbonyl group nucleophilically. (However, again we could turn things round and describe this as electrophilic attack of the $\delta+$ carbon in acetyl chloride on the lone pair of electrons in NH_3.)

If both N and E are organic which description you use is simply a matter of choice.

e.g.

$$(CH_3)_3C^+ + CH_2{=}C(CH_3)_2 \rightarrow CH_3{-}C(CH_3)_2{-}CH_2{-}C^+(CH_3)_2$$

Is this nucleophilic attack of the alkene on the carbocation or electrophilic attack of the carbocation on the alkene? (It is both.)

In this and the next chapter we will divide reactions into electrophilic and nucleophilic categories as they are conventionally done. You must remember though that this division is really quite arbitrary and in all the examples given you should *identify for yourself which is the nucleophile and which the electrophile.*

4.2 Arrows

In organic chemistry an arrow always signifies the direction of movement of electrons. Therefore in an electrophilic–nucleophilic reaction it is always the **nucleophile** which shoots the arrow. *Do not* make the mistake of writing something like this:

[benzene ring ← $^+NO_2$] Incorrect

Although we talk of the electrophilic nitronium ion 'attacking the benzene ring', we must draw the arrow in the other direction to show the flow of electrons:

[benzene ring → $^+NO_2$]

4.3 Electrophilic addition to alkenes (and alkynes)

The regions of high π-electron density in alkenes are subject to attack by **electrophilic** reagents which we can represent in general as:

$$>C{=}C< + E^+ \longrightarrow >\overset{+}{C}{-}C{-}E$$

or sometimes

$$\text{bridged ion: } E^+ \text{ bonded to both carbons of } {-}C{-}C{-}$$

The positively charged intermediate that is produced then usually combines with a **nucleophile** (often the other half of the original reagent):

$$\overset{+}{C}-C-E \quad \text{or} \quad \text{(bridged } E^+ \text{ ion)} \xrightarrow{N^-} N-C-C-E$$

The following are the major reactions which fall into this general category:

(a) Addition of hydrogen halides

$$C{=}C \;\; H{-}I \longrightarrow \overset{+}{C}-C-H + I^- \longrightarrow I-C-C-H$$

(b) Addition of concentrated H_2SO_4

$$C{=}C \;\; H{-}O{-}SO_2{-}OH \longrightarrow \overset{+}{C}-C-H + HSO_4^- \longrightarrow HOSO_2O-C-C-H$$

(c) Addition of water in presence of an acid

$$C{=}C \;\; H^+ \longrightarrow \overset{+}{C}-C-H \xrightarrow{H_2O} H_2\overset{+}{O}-C-C-H \xrightarrow{-H^+} HO-C-C-H$$

(d) Mercuration–demercuration (an alcohol synthesis)

$$C{=}C \;\; Hg(OCOCH_3)_2 \longrightarrow \overset{+}{C}-C-Hg-OCOCH_3 + CH_3COO^-$$

$$\xrightarrow{H_2O} H_2\overset{+}{O}-C-C-HgOCOCH_3 \xrightarrow{\text{(i) } -H^+ \text{ (ii) } NaBH_4} HO-C-C-H$$

(e) Addition of a carbocation [section 1.7(d)]

$$>C{=}C< + R^+ \longrightarrow >\overset{+}{C}-\overset{|}{\underset{|}{C}}-R$$

(another carbocation which can react further and so on)

In all the cases [(a)–(e)], the intermediate is a carbocation. Where there is a choice the more stable carbocation is preferentially formed and the course of the reaction follows Markovnikov's rule. (If you are asked why, for instance, propene is hydrated in acid conditions to give 2-propanol and not 1-propanol, it is a totally unsatisfactory answer to say 'because of Markovnikov's rule'. You must go into the relative stabilities of $CH_3CH_2CH_2^+$ and $CH_3\overset{+}{C}HCH_3$, and how this is explained in terms of the inductive effect of alkyl groups, and so on [see sections 1.8 and 9.3(a)].)

(f) Halogenation

[The sterochemistry of this reaction is covered in section 11.7(a).]

$$>C{=}C< \; Br{-}Br \longrightarrow \text{bromonium ion } (Br^+ \text{ bridging } -C-C-) + Br^-$$

$$\text{bromonium ion} + {}^-Br \longrightarrow -\overset{Br}{\overset{|}{C}}-\underset{Br}{\underset{|}{C}}-$$

Or, in bromine water:

$$\text{bromonium ion} + :OH_2 \longrightarrow -\overset{Br}{\overset{|}{C}}-\underset{^+OH_2}{\underset{|}{C}}- \xrightarrow{-H^+} -\overset{Br}{\overset{|}{C}}-\underset{OH}{\underset{|}{C}}-$$

(a 'halohydrin')

(g) Epoxidation

$$>C{=}C< \; HO{-}O{-}\overset{O}{\overset{\|}{C}}{-}R \longrightarrow -C-C- \text{ (bridged by } \overset{H}{O^+}\text{)} + RCOO^-$$

(a peracid)

$$\xrightarrow{-H^+} -C-C- \text{ (bridged by O, an epoxide)}$$

Notice how similar epoxidation is to formation of the cyclic bromonium ion in the preceding example.

(h) Hydroboration

$$>C{=}C< \quad BH_3 \longrightarrow -\overset{|}{C}{=}{=}\overset{|}{C}- \;(\delta+ \; H{-}BH_2 \; \delta-) \longrightarrow -\overset{|}{\underset{H}{C}}-\overset{|}{\underset{BH_2}{C}}-$$

The hydride ion is transferred more or less simultaneously with formation of the C—B bond. Recall that the orientation of addition is anti-Markovnikov and revise the explanation for this [see section 1.7(c).]

Table 4.1

Reagent	E^+
Hydrogen halides	H^+
H_2SO_4	H^+
H_3O^+/H_2O	H^+
$Hg(OCOCH_3)_2$	$CH_3COO{-}Hg^+$
a protonated alkene	R^+
Br_2	Br^+
Cl_2	Cl^+
RCOO—OH (a peracid)	HO^+
B_2H_6	(BH_3)

In summary, we can list the actual species (E^+) which each of these electrophilic reagents transfers to the double bond. See Table 4.1. Hydroboration is rather an exception in that a positively charged species is not actually added to the double bond. However, the boron atom in BH_3 has only six valence electrons and is therefore able to accept a pair of electrons from the alkene in much the same way as the various E^+ in Table 4.1. (That is, BH_3 is another electrophile.)

(In all the examples of (a)–(h) above I have used an alkene. You should revise the products formed when various electrophiles react with alkynes.)

4.4 Electrophilic aromatic substitution

Like the π-electrons in an alkene or alkyne, the π-electrons in a benzene ring also provide a target for an electron-seeking reagent (or electrophile). Thus we are not surprised that several of the species which we have just seen attacking alkenes (e.g. conc. H_2SO_4, Br_2, R^+) also react with **arenes** (aromatic compounds).

There is certainly a 'connection' then between alkenes, alkynes and arenes in their susceptibility to electrophilic attack. However, there is also an important difference, because:

> electrophiles **add** to alkenes and alkynes, but **substitute** into arenes

The reason for this difference is not hard to find. If an arene were to undergo an addition reaction it would lose the great delocalisation stability associated with the aromatic ring (which can be described in resonance terms or by molecular orbital theory [see section 10.6]).

e.g.

SO$_3$ in conc. H$_2$SO$_4$

addition → H, H, H, OSO$_3$H is **not** formed

(no longer aromatic)

substitution → SO$_3$H + H$_2$O

(still aromatic)

The mechanism of electrophilic substitution involves a two step process and in general is:

E$^+$ → + H E → E + H$^+$

(Do not forget that benzene is C_6H_6. In the intermediate above I have drawn the position of only one of the hydrogens—the important one, since it is replaced by E—but there is of course a hydrogen at each of the other positions around the ring.) We will consider later the positively charged intermediate (and its various other resonance forms and how they can explain reactivity and orientation in the reactions of substituted benzenes [section 10.7]. For the present, let us just list the different sorts of aromatic substitution reaction and in each case note the nature of the actual **electrophile** which attacks the ring.

(a) Nitration

e.g.

$$\text{C}_6\text{H}_6 \xrightarrow[\text{conc. } H_2SO_4]{\text{conc. } HNO_3} \text{C}_6\text{H}_5NO_2$$

The electrophile is the nitronium ion, NO_2^+:

$$\begin{aligned} HO{-}NO_2 + H_2SO_4 &\rightarrow H_2O^+{-}NO_2 + HSO_4^- \\ H_2O^+{-}NO_2 &\rightarrow H_2O + NO_2^+ \\ H_2O + H_2SO_4 &\rightarrow H_3O^+ + HSO_4^- \end{aligned}$$

Overall reaction:

$$HNO_3 + 2H_2SO_4 \rightarrow NO_2^+ + 2HSO_4^- + H_3O^+$$

(Simple salts of the nitronium ion such as $NO_2^+BF_4^-$ and $NO_2^+ClO_4^-$ are known, and are capable of nitrating aromatic compounds.) We could therefore substitute NO_2^+ into the first step of our general mechanism above:

$$\text{C}_6\text{H}_6 + \overset{+}{N}O_2 \longrightarrow \text{C}_6\text{H}_6(H)(NO_2)^+ \rightarrow \text{etc.}$$

(b) Sulphonation

e.g.

$$\text{C}_6\text{H}_6 \xrightarrow{\text{fuming } H_2SO_4} \text{C}_6\text{H}_5SO_3H$$

The electrophile here is sulphur trioxide which is present in fuming H_2SO_4 ('oleum'). The sulphur atom in SO_3 is electrophilic because of polarisation:

$$O^{\delta-}{=}S^{\delta+}({=}O^{\delta-})_2$$

$$\text{C}_6\text{H}_6 + SO_3 \longrightarrow \text{C}_6\text{H}_6(H)(SO_3^-)^+ \longrightarrow \text{C}_6\text{H}_5SO_3H$$

(Note: The intermediate is formed from neutral reactants so it must also be neutral—to counterbalance the positive charge on the benzene ring there is a negative charge on the $-SO_3^-$ group. In the second step the proton to be substituted is lost and another proton is gained by $-SO_3^-$ to give $-SO_3H$, the sulphonic acid group.)

(c) Halogenation

e.g.

$$C_6H_6 + Br_2 \xrightarrow{FeBr_3} C_6H_5Br + HBr$$

In this example, benzene is not brominated in the absence of ferric bromide (unlike an alkene which of course readily reacts with Br_2 without a catalyst.) The function of $FeBr_3$ is to furnish an electrophile more powerful than Br_2 itself; we can think of it as Br^+:

$$Br{-}Br + FeBr_3 \rightarrow Br^{+}{----}FeBr_4^{-}$$

(probably not completely dissociated)

Then:

$$C_6H_6 + Br^+ \longrightarrow C_6H_6Br^+ \rightarrow \text{etc.}$$

(d) Friedel–Crafts alkylation

e.g.

$$C_6H_6 + CH_3CH_2Cl \xrightarrow{AlCl_3} C_6H_5CH_2CH_3 + HCl$$

As we have seen [section 1.2(d)] the electrophile here is a carbocation (again probably not free, but complexed to $AlCl_4^-$):

$$R{-}Cl + AlCl_3 \rightarrow R^{+}{----}AlCl_4^{-}$$

However, it is sufficiently like a free carbocation to undergo rearrangements typical of carbocations (if in so doing it becomes more stable). [See section 1.4.]

(e) Friedel–Crafts acylation

e.g.

$$C_6H_6 + CH_3COCl \xrightarrow{AlCl_3} C_6H_5COCH_3 + HCl$$

The electrophile is an acylium ion:

$$CH_3{-}CO{-}Cl + AlCl_3 \rightarrow CH_3{-}\overset{+}{C}{=}O \quad AlCl_4^- \longleftrightarrow CH_3{-}C{\equiv}O^+ \quad AlCl_4^-$$

(f) Azo dye formation

e.g.

$$C_6H_5{-}\overset{+}{N}{\equiv}N + C_6H_5{-}NMe_2 \longrightarrow C_6H_5{-}N{=}N{-}C_6H_4{-}NMe_2$$

(a diazonium ion) (an azo dye)

The diazonium ion (formed by reaction of a 1° aromatic amine with nitrous acid) is an electrophile, like the other positively charged species we have been looking at in the previous examples.

We can summarise the various types of electrophilic aromatic substitution reactions and the actual species E^+ which is transferred to the benzene ring in the first step of reaction in Table 4.2.

Table 4.2

Reaction	E^+
Nitration	NO_2^+
Halogenation	Cl^+, Br^+
Friedel–Crafts alkylation	R^+
Friedel–Crafts acylation	RCO^+
Azo dye formation	$Ar{-}N_2^+$
(Sulphonation)	(SO_3)

(As we have seen, sulphonation is slightly anomalous in that the attacking species, although an electrophile, is not positively charged.)

4.5 Relative reactivities of electrophiles

Various specific examples of the substitution reactions above may be fast or slow or may not go at all depending on the reactivities of the electrophile involved and of the aromatic compound involved (here acting as a nucleophile). We will consider the latter problem in sections 9.3(b), 9.5(b) and 10.7 (e.g. why aniline is more reactive than benzene, and nitrobenzene less so). At present let us consider the **electrophile**.

For instance, we have seen that bromine is only reactive enough to substitute into benzene when it is complexed with $FeBr_3$, but it is reactive enough on its own to substitute into phenol or aniline.

e.g.

$$C_6H_5NH_2 \xrightarrow[\text{water}]{Br_2} 2,4,6\text{-}Br_3C_6H_2NH_2$$

Likewise dilute nitric acid will nitrate phenol:

$$C_6H_5OH \xrightarrow{HNO_3} o\text{-}O_2NC_6H_4OH + p\text{-}O_2NC_6H_4OH$$

This is because of the presence of some nit*rous* acid, which reacts as follows:

$$HNO_2 + 2HNO_3 \rightleftharpoons H_3O^+ + 2NO_3^- + NO^+$$

The nitrosonium ion (NO^+) is not electrophilic enough to attack benzene but it will substitute into the more reactive phenol. Subsequently, the nitroso substituents (—NO) are rapidly oxidised to nitro groups (—NO_2) by nitric acid (which in turn is reduced to nitrous acid, regenerating it).

The $AlCl_4^-$ complexed carbocation in the Friedel–Crafts reaction is sufficiently electrophilic to react with benzene or even with chlorobenzene, but not with compounds carrying highly deactivating substituents such as —NO_2. Thus, for example, you can nitrate nitrobenzene but you cannot methylate it:

$$C_6H_5NO_2 \xrightarrow[\text{heat}]{\text{conc. } HNO_3/\text{conc. } H_2SO_4} m\text{-}C_6H_4(NO_2)_2$$

$$C_6H_5NO_2 \xrightarrow[AlCl_3]{CH_3Cl} \not\rightarrow \left[m\text{-}CH_3C_6H_4NO_2 \text{ is not formed} \right]$$

Finally, a diazonium ion is an extremely weak electrophile. It will substitute only into a highly activated benzene ring (i.e. one carrying a group such as $-O^-$, [a phenoxide ion], or $-NR_2$). Reactions (a) and (b) are two examples but reaction (c) does *not* occur.

(a) $C_6H_5-\overset{+}{N}\equiv N + C_6H_5-O^-\ Na^+ \longrightarrow C_6H_5-N{=}N-C_6H_4-O^-\ Na^+$

(b)

$^-O_3S-C_6H_4-\overset{+}{N}\equiv N + C_6H_5-NMe_2 \longrightarrow$

$^-O_3S-C_6H_4-N{=}N-C_6H_4-NMe_2$

(Methyl Orange)

(c) $C_6H_5-\overset{+}{N}\equiv N + C_6H_5-CH_3 \not\longrightarrow$

[$C_6H_5-N{=}N-C_6H_4-CH_3$ is not formed]

4.6 Halogenation of aldehydes and ketones

Since we have examined above the electrophilic attack of halogen on alkenes and arenes it seems appropriate to include in this section the halogenation of aldehydes and ketones.

Under **basic** conditions the first step is generation of an enolate ion by removal of an α-hydrogen [see section 2.5]:

$$-\overset{O}{\overset{\|}{C}}-\overset{H}{\overset{|}{C}}- \;\; {}^-OH \longrightarrow -\overset{O}{\overset{\|}{C}}-\overset{-}{C}\lt \longleftrightarrow -\overset{O^-}{\overset{|}{C}}{=}C\lt$$

(main contributor)

And then:

$$-\overset{^-O}{\overset{|}{C}}{=}C\lt \;\; Br{-}Br \rightarrow -\overset{O}{\overset{\|}{C}}-\overset{|}{\underset{|}{C}}-Br + Br^-$$

Once again you have a choice here—do you prefer to think of this as electrophilic attack of bromine on the enolate ion or nucleophilic attack of the enolate ion on bromine?

If there is more than one α-hydrogen then we get more than one bromine substituted into the aldehyde or ketone.

e.g.

$$CH_3CH_2COCH_3 \xrightarrow{Br_2/OH^-} CH_3CH_2COCBr_3$$

(We will consider in section 9.4(a) why this product is formed and not, say, $CH_3CBr_2COCH_3$ or $CH_3CHBrCOCH_2Br$.)

Under **acidic** conditions, the interconversion of keto and enol forms is catalysed and then it is the enol form which is attacked by bromine:

(keto) ⇌ (H⁺) ⇌ (−H⁺) (enol)

$$-C(=O)-CH< \;\underset{}{\overset{H^+}{\rightleftharpoons}}\; -C(={}^{+}OH)-CH< \;\overset{-H^+}{\rightleftharpoons}\; -C(OH)=C<$$

$$-C(\ddot{O}H)=C< \;+\; Br-Br \longrightarrow -C(={}^{+}OH)-C(Br)< \;+\; Br^-$$

$$-C(={}^{+}O-H)-C(Br)< \;\xrightarrow{-H^+}\; -C(=O)-C(Br)<$$

Now the 'connection' I want you to see here is that between the halogenation of an **enol** and the halogenation of a **phenol**:

$$-C(\ddot{O}H)=C< \;+\; Br-Br \longrightarrow -C(={}^{+}OH)-C(Br)< \;+\; Br^-$$

$$C_6H_5\ddot{O}H \;+\; Br-Br \longrightarrow {}^{+}OH{=}C_6H_5(Br)(H) \;+\; Br^-$$

There are several points to note here:

(a) Obviously the mechanisms of these two processes are essentially the *same*.

(b) A phenol, at least when represented by its Kekulé formula, *is* an enol. It is the (stable) enol of the (hypothetical, unstable) ketone:

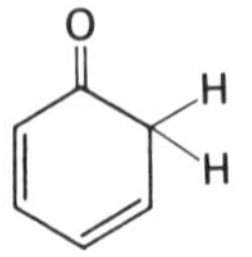

(c) Usually, as we will see in section 10.7, we draw three resonance forms for the intermediate which arises when an electrophile attacks a benzene ring, with the positive charge placed at three positions around the ring. If the ring carries an activating group (such as —OH or $-NH_2$) however, then an extra resonance form with the positive charge taken up by the O or N atom can be drawn and this is the most important contribution to the hybrid (since all atoms have completed valence shells). It is this form that I have drawn above for phenol (for ortho substitution; a similar form can be drawn for para but not for meta substitution [see section 10.7(c)]).

(d) After attack of bromine on either the enol or the phenol the last step is loss of a proton. For the enol the proton on the oxygen atom is lost:

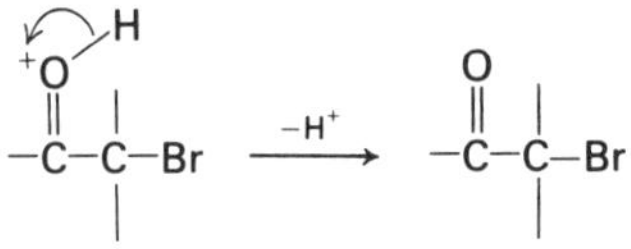

but in the case of the phenol, this would result in loss of the aromaticity of the benzene ring:

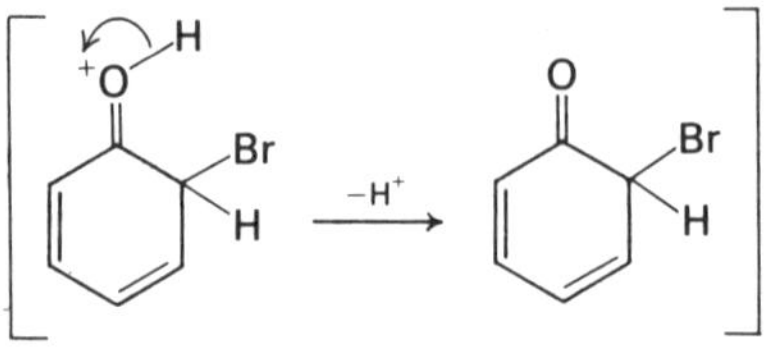

So instead we get loss of the proton from the carbon atom carrying the new bromine substituent and re-formation of the delocalised benzene ring structure:

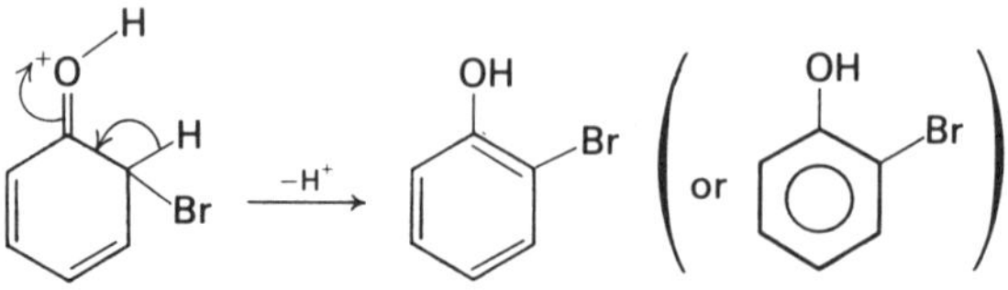

4.7 Hell–Volhard–Zelinsky reaction

This reaction achieves the bromination of a carboxylic acid at the α-position:

e.g.

$$CH_3CH_2CH_2COOH \xrightarrow[Br_2]{\text{red phosphorus}} CH_3CH_2CH(Br)COOH$$

It is appropriate to consider this reaction here because mechanistically it is very similar to the halogenation of aldehydes and ketones which we have just looked at. The first step is formation of the acyl halide brought about by PBr_3 (which is formed from Br_2 and the catalytic amount of phosphorus added):

$$R{-}COOH \xrightarrow{PBr_3} RCOBr$$

The acyl halide is in equilibrium with an enol form (just as is the case with an aldehyde or ketone):

$$Br{-}C(=O){-}C(H)< \;\underset{}{\overset{\pm H^+}{\rightleftharpoons}}\; Br{-}C(OH){=}C<$$

The enol form is electrophilically attacked by bromine (again just as in the halogenation of an aldehyde or ketone):

$$Br{-}C(\ddot{O}H){=}C< \;\; Br{-}Br \longrightarrow Br{-}C(={}^+OH){-}C{-}Br + Br^- \xrightarrow{-H^+} Br{-}C(=O){-}C{-}Br$$

Finally, the following equilibrium regenerates the unsubstituted acyl halide so the overall reaction can continue:

$$Br{-}C(=O){-}C{-}Br + HO{-}C(=O){-}C{-} \rightleftharpoons HO{-}C(=O){-}C{-}Br + Br{-}C(=O){-}C{-}$$

4.8 Combination of free radicals

There is perhaps an exception to our general rule that an electrophile always reacts with a nucleophile and vice versa [section 4.1]. This is the type of reaction which we have seen occurs occasionally as a termination process in a chain reaction.

e.g.

$$CH_3\cdot + CH_3\cdot \rightarrow CH_3CH_3$$

Here both methyl radicals are obviously 'electron-seeking'; they are both striving to complete an octet in their valency shells, and they both achieve their aim when the new C—C bond is formed.

However, this process need not really be an exception to our rule, since an electrophile is defined as a reagent seeking **a pair of electrons** and a free radical is only concerned with gaining a **single** electron. This means that while both a carbocation (for instance) and a free radical are 'electron-seeking', technically speaking only the carbocation is an electrophile. Thus:

$$\cdot C \cdot \quad \cdot C \cdot \longrightarrow -C : C-$$

(each seeks one electron)

$$C^+ \quad :Nuc \longrightarrow -C : Nuc^+$$

seeks a pair of electrons (is an electrophile)

provides a pair of electrons (is a nucleophile)

Once again I recommend you to go through the examples in this chapter and the next identifying the species which are 'seekers of electron-pairs' and those which are 'providers of electron-pairs'.

5

Nucleophilic Attack

5.1 Introduction

I shall begin this chapter with the by now familiar point that when something is attacked by a nucleophile, then by definition it is an electrophile which is attacked. An organic electrophile is a species with a positive or a partial positive charge; therefore these are the types of reactions we will consider under the heading of nucleophilic attack:

1. $\text{Nuc:} \curvearrowright \mathrm{C}^{+}$

2. $\text{Nuc:} \curvearrowright \overset{\delta+}{\mathrm{C}}\text{—X}$

3. $\text{Nuc:} \curvearrowright \overset{\delta+}{\mathrm{C}}\text{=O}$

(1) is the attack of a nucleophile on a carbocation; (2) is the attack on a saturated organic compound with a good leaving group such as Cl^- or H_2O (i.e. an S_N2 reaction). Since X is electron-withdrawing, the carbon is $\delta+$. Reaction (3) is nucleophilic attack on a carbonyl group as found in many compounds such as aldehydes, esters, etc.

5.2 Repetitiveness

Some of the reactions we will look at in this chapter have already been met. For instance we have studied carbanions, and carbanions often participate in nucleophilic attack. I make no apology for this kind of repetition; it is important to study a given reaction from several points of view. This makes it easier to understand and to see how it fits into the rest of organic chemistry. Take the **aldol condensation**, for example—this comes up in the following sections:

Carbanions It involves a resonance-stabilised carbanion or enolate ion.
Nucleophilic attack The enolate ion attacks the carbonyl group nucleophilically.
Carbon-carbon bond making It is an important and versatile synthetic procedure.

Acid–base reactions It involves loss of an acidic α-hydrogen to give the enolate ion.
Resonance The structure of the enolate ion is described by two resonance forms.

Thus we can see how the aldol condensation is relevant in many areas of organic chemistry. Studying a reaction such as this complete with all its 'connections' is much more fruitful than just looking at it once in isolation.

5.3 Nucleophilic attack on a carbocation

As we have already seen [section 1.2] one of the main reactions of carbocations is combination with a nucleophile. An example of this is a classical S_N1 reaction such as the following reaction of *t*-butyl bromide (2-bromo-2-methylpropane) with silver nitrate solution, where water is the nucleophile:

$$(CH_3)_3C{-}Br \rightleftharpoons CH_3{-}\overset{+}{C}(CH_3)_2 + Br^- \xrightarrow{Ag^+} AgBr\downarrow$$

$$H_2\ddot{O}\ \ CH_3{-}\overset{+}{C}(CH_3)_2 \longrightarrow H_2\overset{+}{O}{-}C(CH_3)_2{-}CH_3 \xrightarrow{-H^+} HO{-}C(CH_3)_2{-}CH_3$$

This sort of process is characteristic of a *tertiary* alkyl halide; ionisation is relatively favourable (a) because the 3° carbocation is stabilised by the inductive effect of the three alkyl groups and (b) because of the relief of steric crowding of four bulky groups round the carbon atom in the original alkyl halide.

S_N1 = substitution, nucleophilic, unimolecular. The rate determining step (the slowest step) is the first one, the dissociation of alkyl halide, which is a unimolecular process.

Other types of alkyl halide which can react in the same way are allylic and benzylic halides [section 1.9]. These give rise to resonance-stabilised carbocations which again can be attacked by nucleophiles.

e.g.

$$CH_2{=}CH{-}CH_2Br \rightleftharpoons CH_2{=}CH{-}CH_2^+ + Br^-\ (\rightarrow AgBr)$$

$$\updownarrow$$

$$^+CH_2{-}CH{=}CH_2$$

$$H_2O\colon\ \overset{+}{C}H_2{-}CH{=}CH_2 \longrightarrow H_2\overset{+}{O}{-}CH_2{-}CH{=}CH_2 \xrightarrow{-H^+} HOCH_2{-}CH{=}CH_2$$

The first reaction in this section—the hydrolysis of *t*-butyl bromide (2-bromo-2-methylpropane)—is different of course from the acid-catalysed hydration of isobutene (methylpropene). Or is it? No, in fact the mechanism involves *exactly the same step* in both cases, namely the nucleophilic attack of water on the *t*-butyl carbocation:

$$(CH_3)_2C{=}CH_2 \xrightarrow{H^+} (CH_3)_2C^+{-}CH_3$$

$$H_2O{:} \; (CH_3)_2C^+{-}CH_3 \longrightarrow H_2\overset{+}{O}{-}C(CH_3)_3 \xrightarrow{-H^+} HO{-}C(CH_3)_3$$

Furthermore, the acid-catalysed hydrolysis of *t*-butyl acetate involves again *exactly the same step*. [See if you can write the mechanism before checking back to section 1.7(a).]

Water is not the only nucleophile which can attack a carbocation. In the addition of H—Cl and H_2SO_4 to alkenes the first step is protonation of the alkene to give a carbocation, which then combines with the nucleophile Cl^- or HSO_4^- [section 1.2(b)].

e.g.

$$CH_3CH{=}CH_2 \xrightarrow{H_2SO_4} CH_3{-}\overset{+}{C}H{-}CH_3$$

$$HO{-}S(=O)_2{-}O^- \;\; CH_3{-}\overset{+}{C}H{-}CH_3 \longrightarrow (CH_3)_2CH{-}OSO_3H$$

The chloride ion is also the nucleophile when an alcohol reacts with the Lucas reagent ($ZnCl_2$/conc. HCl).

e.g.

$$CH_3{-}C(CH_3)_2{-}OH + ZnCl_2 \longrightarrow CH_3{-}C(CH_3)_2{-}\overset{\delta+}{O}(H){-}{-}\overset{\delta-}{Z}nCl_2 \rightleftharpoons CH_3{-}{}^{+}C(CH_3)_2 + [ZnCl_2OH]^-$$

Then:

$$Cl^- \;\; CH_3{-}\overset{+}{C}(CH_3)_2 \longrightarrow Cl{-}C(CH_3)_3$$

We have looked at Friedel–Crafts alkylation in the chapter on electrophilic attack [section 4.4(d)] but we could equally well think of it as nucleophilic attack of the **arene** on a carbocation (complexed to $AlCl_4^-$).

e.g.

$$C_6H_6 + CH_3\overset{+}{C}H(CH_3)\cdots AlCl_4^- \longrightarrow \overset{+}{C_6H_6}(H)CH(CH_3)_2 \rightarrow \text{etc.}$$

As a last example consider treating *t*-butyl bromide with $AgNO_3$ in ethanol solution (rather than water). This involves nucleophilic attack of ethanol on the carbocation to give an ether (rather than an alcohol):

$$CH_3CH_2\ddot{O}H + (CH_3)_3C^+ \longrightarrow CH_3CH_2{-}\overset{+}{O}(H){-}C(CH_3)_3 \xrightarrow{-H^+} CH_3CH_2{-}O{-}C(CH_3)_3$$

Other methods of producing the *t*-butyl carbocation in ethanol could also lead to the same product (e.g. *t*-butanol + acid + ethanol, or isobutene + acid + ethanol). It could also be made by the Williamson ether synthesis utilising one of these routes, but not the other:

$$CH_3CH_2O^-Na^+ + (CH_3)_3C{-}Br \rightarrow ?$$

$$(CH_3)_3C{-}O^-Na^+ + CH_3CH_2Br \rightarrow ?$$

Which one? See section 7.7(b).

5.4 Nucleophilic attack at a saturated carbon

Just as the previous section was typified by the S_N1 mechanism so this section is most characteristically represented by the classical S_N2 process. In general (neglecting charges):

$$Nuc{:} + CXYZ{-}LG \longrightarrow Nuc{-}CXYZ + LG$$

The nucleophile attacks at the 'back-side' of the molecule pushing out the leaving group (LG) and inverting the configuration at the carbon atom. (Inversion is only apparent of course if we are dealing with optically active substances.) Two molecules are involved in the rate determining step (in fact the only step of the reaction)—hence S_N2.

S_N2 processes occur with 1° and 2° compounds. If X, Y and Z are all alkyl groups the attack of the nucleophile is sterically hindered and then either the S_N1 mechanism comes into play or elimination takes over instead of substitution. [Elimination is considered in section 7.7.]

Let us now list some examples of S_N2 reactions. In every case you should appreciate how the reaction is 'connected' to the others by conforming to the general reaction mechanism shown above.

(a) Interconversion of alkyl halides

e.g.

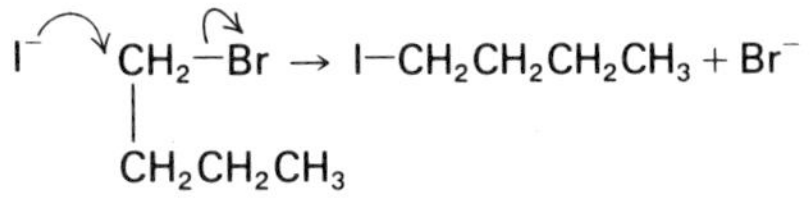

$$I^- + CH_2(CH_2CH_2CH_3){-}Br \rightarrow I{-}CH_2CH_2CH_2CH_3 + Br^-$$

(b) Hydrolysis of alkyl halides

e.g.

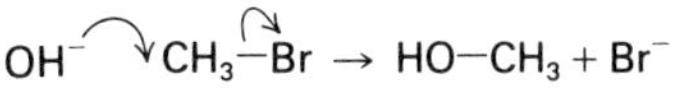

$$OH^- + CH_3{-}Br \rightarrow HO{-}CH_3 + Br^-$$

(c) Williamson ether synthesis (reaction of an alkyl halide with the sodium salt of an alcohol or sodium salt of a phenol)

e.g.

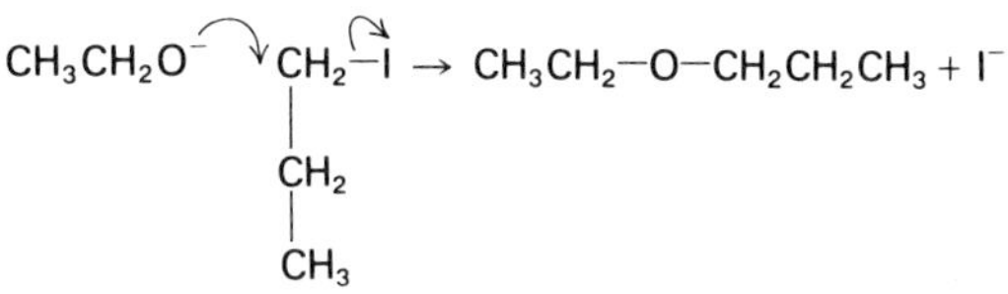

$$CH_3CH_2O^- + CH_2(CH_2CH_3){-}I \rightarrow CH_3CH_2{-}O{-}CH_2CH_2CH_3 + I^-$$

The following is an example found in the synthetic route to the pain-killer, phenacetin:

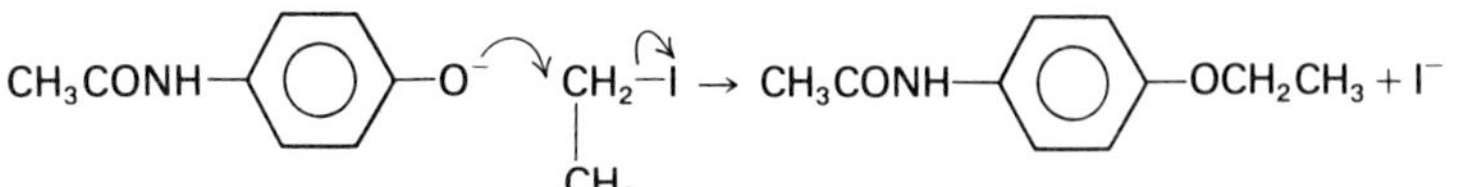

$$CH_3CONH{-}C_6H_4{-}O^- + CH_2(CH_3){-}I \rightarrow CH_3CONH{-}C_6H_4{-}OCH_2CH_3 + I^-$$

phenacetin or
N-(4-ethoxyphenyl)ethanamide

(d) Epoxide formation

$$\begin{array}{c} \mathrm{Br} \\ | \\ -\mathrm{C}-\mathrm{C}- \\ | \\ \mathrm{OH} \end{array}$$

a 'halohydrin'—how is this made? [section 4.3(f)]

NaOH ⇌

$$-\mathrm{C}(\mathrm{O}^-)-\mathrm{C}(\mathrm{Br})- \longrightarrow -\mathrm{C}\!-\!\mathrm{C}-\ (\text{epoxide ring with O}) + \mathrm{Br}^-$$

This is essentially an internal Williamson ether synthesis, the nucleophilic attack of an alkoxide ion on an alkyl halide within the same molecule. It is like an S_N2 mechanism, but is given the designation S_Ni, where *i* denotes an intramolecular reaction. How else may epoxides be made? [See section 4.3(g).]

(e) Reaction of alkyl halides with ammonia

e.g.

$$\ddot{\mathrm{N}}\mathrm{H}_3\ \ \mathrm{CH_3CH_2{-}Br} \longrightarrow \mathrm{H_3\overset{+}{N}{-}CH_2CH_3} + \mathrm{Br}^-$$

$$\downarrow -\mathrm{H}^+$$

$$\mathrm{H_2N{-}CH_2CH_3}$$

(f) Reaction of alkyl halides with amines

e.g.

$$\mathrm{CH_3\ddot{N}H_2}\ \ \mathrm{CH_3{-}Cl} \longrightarrow \mathrm{CH_3{-}\overset{+}{N}H_2{-}CH_3} + \mathrm{Cl}^-$$

$$\downarrow -\mathrm{H}^+$$

$$\mathrm{CH_3{-}NH{-}CH_3}$$

$$\mathrm{(CH_3CH_2)_2\ddot{N}H}\ \ \mathrm{CH_3{-}I} \longrightarrow \mathrm{(CH_3CH_2)_2\overset{+}{N}H{-}CH_3} + \mathrm{I}^-$$

$$\downarrow -\mathrm{H}^+$$

$$\mathrm{Et_2NMe}$$

(g) Reaction of alkyl halides with the sodium salt of phthalimide

e.g.

$$C_6H_4(CO)_2N^- + CH_2(CH_2CH_2CH_3){-}Br \longrightarrow C_6H_4(CO)_2N{-}CH_2CH_2CH_2CH_3 + Br^-$$

The product can be hydrolysed to the primary amine—in this case, 1-butanamine [see section 5.5(u)]. The whole process is the Gabriel 1°-amine synthesis.

(h) Reaction of alkyl halides with cyanide ion

e.g.

$$N{\equiv}C^- + CH_2(CH_2OH){-}Cl \rightarrow HOCH_2CH_2CN + Cl^-$$

(i) Reaction of alkyl halides with acetylide (alkynide) ions

e.g.

$$CH_3CH_2C{\equiv}C^- + CH_2(CH_3){-}Br \rightarrow CH_3CH_2C{\equiv}CCH_2CH_3 + Br^-$$

(j) Reaction of alkyl halides with $LiAlH_4$

e.g.

$$H^- + CH_3{-}Br \rightarrow CH_4 + Br^-$$

(from $LiAlH_4$)

($LiAlH_4$ and $NaBH_4$ may be thought of essentially as providers of hydride ions, H^-.)

(k) Reaction of alkyl halides with enolate ions

e.g. the resonance-stabilised carbanion derived from ethyl acetoacetate (ethyl 3-oxobutanoate) + base, or diethyl malonate (diethyl propanedioate) + base:

$$CH_3-\overset{\overset{O}{\|}}{C}-\overset{-}{C}H-COOEt + CH_3-Br \rightarrow CH_3-\overset{\overset{O}{\|}}{C}-\overset{\overset{CH_3}{|}}{C}H-COOEt + Br^-$$

$$(EtOOC)_2CH^- \;\; CH_2(CH_2CH_3)-I \longrightarrow CH_3CH_2CH_2CH(COOEt)_2 + I^-$$

(l) Corey–House alkane synthesis

e.g.

$$(CH_3)_2CuLi + CH_3(CH_2)_4-I \rightarrow CH_3(CH_2)_4-CH_3$$

The alkyl groups in a lithium dialkyl cuprate may be considered as 'carbanion-like' as in a Grignard or other organometallic reagent, so essentially we have here the following type of mechanism which is clearly the same as all the others in this section:

$$R^- \;\; R'-X \rightarrow R-R' + X^-$$

(m) Conversion of alcohols to alkyl halides

This might be achieved using dry hydrogen bromide, or sodium bromide and concentrated sulphuric acid.

e.g.

$$CH_3CH_2CH_2CH_2OH \overset{H^+}{\rightleftharpoons} CH_3CH_2CH_2CH_2-\overset{+}{O}H_2$$

$$Br^- \;\; CH_2(CH_2CH_2CH_3)-\overset{+}{O}H_2 \longrightarrow Br-CH_2CH_2CH_2CH_3 + H_2O$$

Note that without acid, the bromide ion will not react with the alcohol:

$$Br^- \;\; R-OH \not\longrightarrow R-Br + OH^-$$

The hydroxide ion is a poor leaving group; the function of the acid is to protonate the alcohol so that H_2O (a better leaving group) departs rather than OH^-. [Leaving groups feature in Chapter 8.]

(n) Conversion of alcohols to ethers

In the acid-catalysed dehydration of alcohols to ethers we get an exactly analogous process to the previous example:

$$CH_3CH_2\ddot{O}H \curvearrowright \underset{\displaystyle CH_3}{\underset{|}{CH_2}}{-}\overset{+}{O}H_2 \longrightarrow CH_3CH_2{-}\underset{\displaystyle H}{\underset{|}{\overset{+}{O}}}{-}CH_2CH_3 + H_2O$$

$$CH_3CH_2OH \xrightleftharpoons{H^+} \text{(protonated alcohol)} \qquad \xrightarrow{-H^+} Et_2O$$

An unprotonated alcohol molecule nucleophilically attacks a protonated one. Again, acid is necessary to convert —OH to a better leaving group.

(o) Cleavage of ethers

Just like an alcohol, an ether will not react with halide ion until it is protonated. Then:

$$I^- \curvearrowright R{-}\underset{\displaystyle H}{\underset{|}{\overset{+}{O}}}{-}R' \longrightarrow R{-}I + R'{-}OH$$

$$R{-}O{-}R' \xrightleftharpoons{H^+} R{-}\overset{+}{O}H{-}R' \qquad R'{-}OH \xrightarrow{\text{[reacts with HI as in (m)]}} R'{-}I$$

In other words, just as H_2O is a better leaving group than OH^-, so is $R'{-}OH$ a better leaving group than $R'{-}O^-$.

(p) Nucleophilic attack on epoxides

The previous example is about the only one involving nucleophilic substitution of an ordinary ether. A three membered cyclic ether, however, is very much more reactive (because of ring-strain) and can be opened by the nucleophilic attack of various reagents.

e.g.

$$HO^- + CH_2{-}CH_2 \text{ (epoxide, O)} \longrightarrow HO{-}CH_2{-}CH_2{-}O^- \xrightarrow{H_2O} HOCH_2CH_2OH$$

$$H_3\ddot{N} + CH_2{-}CH_2 \text{ (epoxide, O)} \longrightarrow H_3\overset{+}{N}{-}CH_2CH_2{-}O^- \xrightarrow{\pm H^+} H_2N{-}CH_2CH_2{-}OH$$

$$R^- \text{ (RMgBr)} + CH_2{-}CH_2 \text{ (epoxide, O)} \longrightarrow R{-}CH_2CH_2{-}O^- \xrightarrow{H_2O} R{-}CH_2CH_2OH$$

Hydrolysis of an epoxide can occur under acidic conditions too (but not the other reactions shown—acid would simply protonate the nucleophiles, NH_3 and R^-). Acid-catalysed hydrolysis of an epoxide follows the mechanism:

$$\text{epoxide} \xrightleftharpoons{H^+} \text{protonated epoxide } (O^+{-}H) \xrightarrow{H_2O:} H_2O^+{-}C{-}C{-}OH \xrightarrow{-H^+} HO{-}C{-}C{-}OH$$

Note the 'connection' between this mechanism and that of bromination of an alkene [section 4.3(f)]. After electrophilic attack of Br_2 we get nucleophilic attack of Br^- on the cyclic bromonium ion (which is just like nucleophilic attack of H_2O on the cyclic oxonium ion above):

$$Br^- + \text{cyclic bromonium ion } (Br^+) \longrightarrow Br{-}C{-}C{-}Br$$

and 'bromohydrin' formation [alkene + bromine water, see also section 11.5] involves the even more similar process:

$$H_2O: + \text{cyclic bromonium ion } (Br^+) \longrightarrow H_2O^+{-}C{-}C{-}Br \xrightarrow{-H^+} HO{-}C{-}C{-}Br$$

(q) Résumé

To reiterate, all the numerous examples in this section 5.4 are strongly connected; they all share essentially the same mechanism which we have generalised as:

$$\text{Nuc:} \curvearrowright -\overset{|}{\underset{|}{C}}{}^{\delta+}\!-\!X$$

5.5 Nucleophilic attack on acyl derivatives

In this section we will look at nucleophilic attack on all those compounds containing carbonyl groups *except aldehydes and ketones* (that is, carboxylic acids, acyl chlorides, anhydrides, esters, amides and imides).

There are some 'connections' here to those S_N2 reactions we have just been looking at, for example in the similarity between simple *equations* for reactions such as:

$$NH_3 + CH_3CH_2Cl \rightarrow CH_3CH_2NH_2 + HCl$$

$$NH_3 + CH_3COCl \rightarrow CH_3CONH_2 + HCl$$

And both processes of course involve attack of a nucleophile:

$$H_3\ddot{N} \curvearrowright CH_2{-}Cl \ (\text{with } CH_3 \text{ on } CH_2) \qquad H_3\ddot{N} \curvearrowright C{=}O \ (\text{with } Cl \text{ and } CH_3 \text{ on } C)$$

However, there is an important difference too. **S_N2 reactions have a single step mechanism whereas nucleophilic substitution at a carbonyl group is a two step process**. We can show this clearly with an energy profile for the progress of each of the reactions. For an S_N2 process, this is given by Fig. 5.1.

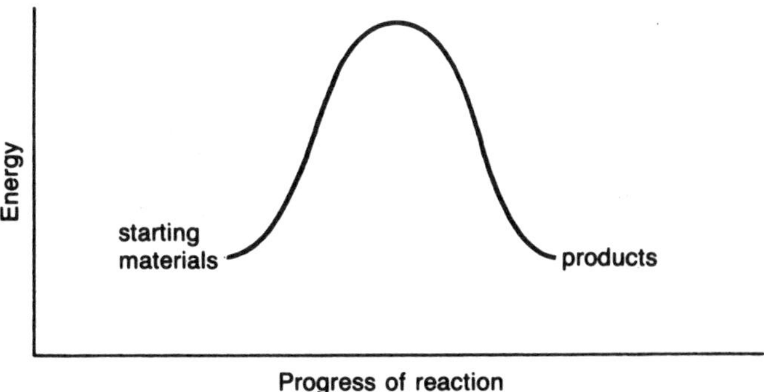

Fig. 5.1

Ascending the left hand edge of the energy barrier, the bond between the nucleophile and carbon is starting to form and that between the leaving group and carbon is starting to break. The **transition state** (top of the energy barrier) could be represented by:

Nuc----C----LG

partly formed   partly broken

Down the right hand edge the bond between the nucleophile and carbon gets stronger and stronger, whereas that between carbon and the leaving group gets weaker and weaker until finally it is non-existent and the leaving group is free to leave. The whole process is a single concerted action; as the nucleophile approaches, the leaving group departs.

However, a different reaction profile occurs for displacement at acyl carbon, see Fig. 5.2. The **first step** involves attack of the nucleophile to give a **tetrahedral intermediate** (so called as there are now four bonds to carbon where before, in the acyl compound, the coordination was trigonal).

$$\text{Nuc}^- + \text{X(R)C{=}O} \rightarrow \text{Nuc{-}C(X)(R){-}O}^-$$

This intermediate is a real chemical species which may be more or less stable and more or less long-lived; it is not the same as a transition state which is an energy maximum with no finite lifetime.

Subsequently, in a **second step**, the intermediate breaks down to re-form a trigonal carbonyl containing species and eliminate the leaving group:

$$\text{Nuc{-}C(X)(R){-}O}^- \rightarrow \text{Nuc(R)C{=}O} + \text{X}^-$$

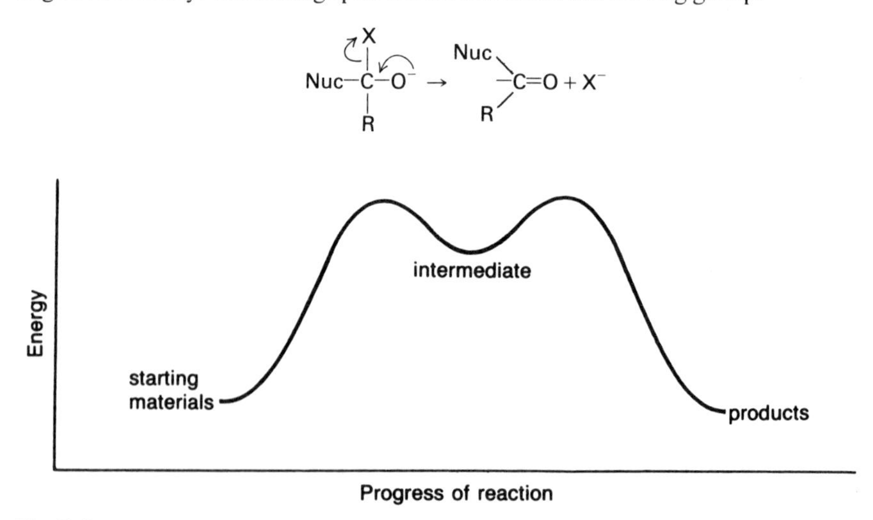

Fig. 5.2

Now we will list examples of nucleophilic displacement reactions involving acyl compounds—once again, concentrate on recognising the way in which all these reactions share essentially the same mechanism. (They all have the two steps we have looked at above; sometimes extra steps are also involved such as the final loss of a proton, or perhaps the initial protonation of the carbonyl group.)

(a) Hydrolysis of acyl halides

e.g.

$$H_2\ddot{O}: + CH_3-C(Cl)=O \rightarrow H_2\overset{+}{O}-C(Cl)(CH_3)-O^- \rightarrow H_2\overset{+}{O}-C(CH_3)=O + Cl^- \xrightarrow{-H^+} CH_3COOH$$

Note that the carbon atom which is attacked is doubly attractive to a nucleophile because of polarisation of the carbonyl group and because of the inductive effect of chlorine [see section 9.5(a)]:

$$R-\overset{\delta+}{C}(=O^{\delta-})\rightarrow Cl$$

Therefore an acyl halide is readily hydrolysed by water, whereas an alkyl halide [section 5.4(b)] requires the stronger nucleophile OH^- rather than H_2O itself.

(b) Reaction of acyl halides with alcohols

e.g.

$$CH_3CH_2\ddot{O}H + CH_3-C(Cl)=O \rightarrow CH_3CH_2-\overset{+}{O}(H)-C(Cl)(CH_3)-O^- \rightarrow CH_3CH_2-\overset{+}{O}(H)-C(CH_3)=O + Cl^- \xrightarrow{-H^+} CH_3COOCH_2CH_3$$

This is a standard ester preparation. Note that, as in the last example, an alkyl halide is less reactive—it reacts with R—O⁻ (in the Williamson synthesis) but not R—OH.

(c) Reaction of acyl halides with phenoxide ions

A phenol is less nucleophilic than an alcohol [for the explanation see section 10.2(b)]. Therefore to make the ester of a phenol we do the reaction under basic conditions (the 'Schotten–Baumann reaction') where the stronger nucleophile, the phenoxide ion, is involved.

e.g.

$$C_6H_5O^- + C_6H_5COCl \longrightarrow C_6H_5O{-}C(Cl)(C_6H_5){-}O^- \longrightarrow C_6H_5O{-}C(C_6H_5){=}O + Cl^-$$

(phenyl benzoate)

$$C_6H_5OH \xrightleftharpoons{OH^-} C_6H_5O^-$$

(d) Reaction of acyl halides with ammonia

e.g.

$$\ddot{N}H_3 + C_6H_5COBr \longrightarrow H_3\overset{+}{N}{-}C(Br)(C_6H_5){-}O^- \longrightarrow H_3\overset{+}{N}{-}C(C_6H_5){=}O + Br^- \xrightarrow{-H^+} C_6H_5CONH_2$$

This is a standard **amide** preparation.

(e) Reaction of acyl halides with amines

e.g.

$$CH_3\ddot{N}H_2 + CH_3COCl \rightarrow CH_3{-}\overset{+}{N}H_2{-}C(Cl)(CH_3){-}O^- \rightarrow CH_3{-}\overset{+}{N}H_2{-}C(CH_3){=}O + Cl^- \xrightarrow{-H^+} CH_3CONHCH_3$$

$$(CH_3CH_2)_2\ddot{N}H + PhC(Cl)=O \longrightarrow (CH_3CH_2)_2\overset{+}{N}H-C(Cl)(Ph)-O^- \longrightarrow Et_2\overset{+}{N}H-C(Ph)=O + Cl^- \xrightarrow{-H^+} PhCONEt_2$$

(f) Reaction of acyl halides with carboxylate anions

e.g.

$$CH_3CH_2CH_2COO^- + CH_3C(Cl)=O \longrightarrow CH_3CH_2CH_2-\overset{O}{\overset{\|}{C}}-O-C(Cl)(CH_3)-O^- \longrightarrow CH_3CH_2CH_2-\overset{O}{\overset{\|}{C}}-O-C(CH_3)=O + Cl^-$$

This is an **anhydride** preparation.

(g) Reaction of acyl halides with organocadmium reagents

e.g.

$$(CH_3)_2CHCH_2^- + CH_3CH_2C(Cl)=O \rightarrow (CH_3)_2CHCH_2-C(Cl)(CH_2CH_3)-O^- \rightarrow (CH_3)_2CHCH_2-C(CH_2CH_3)=O + Cl^-$$

(R^- from R_2Cd)

We think of organometallic compounds as sources of carbanions, although as discussed before, an organocadmium has much covalent character and, unlike a Grignard reagent, is not reactive enough to attack the product ketone [see sections 2.7(c), 6.5(a)].

(h) Reaction of acyl halides with lithium tri-*t*-butoxyaluminium hydride

$Li(t\text{-}BuO)_3AlH$ (like $LiAlH_4$ and $NaBH_4$) may be considered as a source of H^-, the hydride ion, which can nucleophilically attack an acyl halide. (Hydride ion should not be considered as being free in this reaction, although it is convenient to write the mechanism as if it were; in fact, it is transferred from the reagent to the carbonyl group.)

$$[(t\text{-BuO})_3AlH]^- \quad H^- + C_6H_5\text{–}C(Cl)\text{=O} \longrightarrow H\text{–}C(Cl)(C_6H_5)\text{–}O^- \longrightarrow H\text{–}C(C_6H_5)\text{=O} + Cl^-$$

The bulky reagent is less reactive than the smaller AlH_4^-. It will not attack the product aldehyde whereas $LiAlH_4$ would do so [section 5.6(g)].

(i) Reactions of anhydrides

In general anhydrides do everything that acyl halides do, but less vigorously as $RCOO^-$ is not such a good leaving group as Cl^-.

e.g.

$$H_2O{:} + CH_3\text{–}C(\text{=O})\text{–O–}C(\text{=O})\text{–}CH_3 \longrightarrow H_2\overset{+}{O}\text{–}C(CH_3)(O^-)\text{–O–CO–}CH_3 \longrightarrow H_2\overset{+}{O}\text{–}C(CH_3)\text{=O} + CH_3COO^-$$

$$H_2\overset{+}{O}\text{–}C(CH_3)\text{=O} \xrightarrow{-H^+} CH_3COOH; \quad CH_3COO^- \xrightarrow{+H^+} CH_3COOH$$

With ethanol and ammonia I will give you the simple equation and you should write out the details of the mechanism which by now should be familiar:

$$CH_3CH_2OH + (CH_3CO)_2O \rightarrow CH_3COOCH_2CH_3 + CH_3COOH$$

$$NH_3 + (CH_3CO)_2O \rightarrow CH_3CONH_2 + CH_3COOH$$

Here is an example, which like the reaction in section 5.4(c), is found in the synthesis of phenacetin:

$$HO\text{–}C_6H_4\text{–}\ddot{N}H_2 + CH_3\text{–}C(\text{=O})\text{–O–CO–}CH_3 \rightarrow HO\text{–}C_6H_4\text{–}\overset{+}{N}H_2\text{–}C(CH_3)(O^-)\text{–O–}COCH_3 \rightarrow HO\text{–}C_6H_4\text{–}\overset{+}{N}H_2\text{–}C(CH_3)\text{=O} + CH_3COO^-$$

$$HO\text{–}C_6H_4\text{–}\overset{+}{N}H_2\text{–}C(CH_3)\text{=O} \xrightarrow{-H^+} HO\text{–}C_6H_4\text{–}NHCOCH_3$$

(j) Reaction of carboxylic acids with lithium aluminium hydride

$LiAlH_4$ is the only reducing agent powerful enough to reduce a carboxylic acid. We can picture the reaction as nucleophilic attack of the powerful nucleophile H^- on $RCOO^-$. First:

$$RCOOH + AlH_4^- \rightarrow RCOO^- + AlH_3 + H_2$$

Then:

$$H^- \text{(from } AlH_4^-) + RC(O^-)=O \longrightarrow H{-}C(O^-)(R){-}O^- \longrightarrow H{-}C(R){=}O + O^{2-} \text{ (not free)}$$

(The oxide ion ends up complexed in an aluminium salt.) The aldehyde is immediately reduced further (to the 1° alcohol, RCH_2OH) as in section 5.6(g).

(k) Reaction of carboxylic acids with alcohols

By analogy with the example in section 5.6(b) involving an acyl halide and an alcohol we *could* propose the following mechanism:

$$CH_3CH_2\ddot{O}H + CH_3C(OH)=O \longrightarrow CH_3CH_2{-}\overset{+}{O}(H){-}C(OH)(CH_3){-}O^- \longrightarrow CH_3CH_2{-}\overset{+}{O}(H){-}C(CH_3){=}O + OH^-$$

$$\xrightarrow{-H^+} CH_3COOCH_2CH_3$$

There are two reasons why this process does not happen (or at least not at an appreciable rate). Firstly, the δ+ charge in the polarised carbonyl group is partly offset by the oxygen atom next to it:

$$R{-}\overset{\delta+}{C}(=O^{\delta-}){-}\ddot{O}H \longleftrightarrow R{-}C(O^-){=}\overset{+}{O}H$$

This means that the carbonyl group is not so readily attacked by the nucleophile (ethanol). Secondly, the leaving group in the process above is OH^-, a moderately strong base and therefore a poor leaving group.

To get round both these problems we carry out the reaction under acidic conditions. Now, to some extent at least, the carboxyl group is protonated; it carries a full positive charge and is thus rendered more attractive to the nucleophile. Furthermore, under these conditions it is H_2O that leaves rather than the poorer leaving group, OH^-.

$$CH_3COOH \xrightleftharpoons{H^+} CH_3CH_2\ddot{O}H + CH_3\overset{+}{C}(OH)=OH \rightleftharpoons CH_3CH_2\overset{+}{O}(H){-}C(OH)(CH_3){-}OH \xrightleftharpoons{\pm H^+} CH_3CH_2{-}O{-}C(\overset{+}{O}H_2)(CH_3){-}OH \rightleftharpoons CH_3CH_2{-}O{-}C(CH_3)=\overset{+}{O}H + H_2O \xrightleftharpoons{-H^+} CH_3COOCH_2CH_3$$

Note: H^+ is not consumed—it is only acting as a catalyst. All these transformations are reversible equilibria; to get a good yield of ester we would use an excess of one of the starting materials, carboxylic acid or alcohol.

(l) Acid-catalysed ester hydrolysis

This process is *exactly the reverse* of the one we have just looked at, that is:

$$CH_3COOCH_2CH_3 \xrightleftharpoons{H^+} H_2\ddot{O} + CH_3C(OCH_2CH_3)=\overset{+}{O}H \rightleftharpoons H_2\overset{+}{O}{-}C(OCH_2CH_3)(CH_3){-}OH \xrightleftharpoons{\pm H^+} H{-}\overset{+}{O}(CH_2CH_3){-}C(OH)(CH_3){-}OH \rightleftharpoons HO{-}C(CH_3)=\overset{+}{O}H + CH_3CH_2OH \xrightleftharpoons{-H^+} CH_3COOH$$

Carrying out the hydrolysis of esters such as the above in $H_2{}^{18}O$ gives ^{18}O-labelled carboxylic acid but unlabelled alcohol as the products which supports the mechanism shown. This is in contrast to what happens with *t*-butyl esters [section 1.7(a)] which follow a different mechanism in acid-catalysed hydrolysis.

(m) Acid-catalysed transesterification

This process is *exactly the same* as the previous one, except that ROH is the nucleophile instead of H_2O.

e.g.

$$CH_3CO OCH_2CH_3 \underset{}{\overset{H^+}{\rightleftharpoons}} CH_3C(OCH_2CH_3)=\overset{+}{O}H \rightleftharpoons CH_3\text{–}\overset{+}{O}(H)\text{–}C(CH_3)(OCH_2CH_3)\text{–}OH \overset{\pm H^+}{\rightleftharpoons} CH_3\text{–}O\text{–}C(CH_3)(\overset{+}{O}(H)CH_2CH_3)\text{–}OH$$

$$\rightleftharpoons CH_3O\text{–}C(CH_3)=\overset{+}{O}H + CH_3CH_2OH \overset{-H^+}{\rightleftharpoons} CH_3COOCH_3$$

Again, the function of the H^+ is to render the carbonyl group more susceptible to nucleophilic attack, and to ensure that the leaving group is ROH rather than RO^-.

(n) Base-promoted ester hydrolysis

e.g.

$$HO^- + C(OCH_3)(CH_2CH_3)=O \rightleftharpoons HO\text{–}C(OCH_3)(CH_2CH_3)\text{–}O^- \rightleftharpoons HO\text{–}C(CH_2CH_3)=O + CH_3O^-$$

$$HO\text{–}C(CH_2CH_3)=O \xrightarrow{-H^+} CH_3CH_2COO^- \qquad CH_3O^- \xrightarrow{+H^+} CH_3OH$$

Note: (a) this is *not base-catalysed*, because OH^- is consumed in the reaction, and (b) the leaving group is CH_3O^-, a moderately strong base and therefore a poor leaving group. However, it is able to leave since in the next step it immediately gains a proton at the expense of the acidic RCOOH, which ends up as the resonance-stabilised $RCOO^-$ ion [section 10.2(a)]. The stability of $RCOO^-$ furnishes a driving force for the reaction to proceed to completion. If we want RCOOH as the product, the reaction mixture must then be acidified.

(o) Base-catalysed transesterification

e.g.

$$CH_3O^- + CH_3-C(=O)-OCH_2CH_3 \rightleftharpoons CH_3O-C(CH_3)(O^-)-OCH_2CH_3 \rightleftharpoons CH_3O-C(CH_3)=O + CH_3CH_2O^-$$

In excess CH_3OH only a trace of CH_3O^- is required for the reaction to proceed in the forward direction as CH_3O^- is replenished from this equilibrium:

$$\underset{\text{(excess)}}{CH_3OH} + CH_3CH_2O^- \rightleftharpoons CH_3O^- + CH_3CH_2OH$$

(p) The Claisen condensation

The first step is removal of an α-hydrogen to give a resonance-stabilised carbanion.

e.g.

$$CH_3-C(=O)-OEt \underset{}{\overset{EtO^-}{\rightleftharpoons}} {}^-CH_2-C(=O)-OEt \longleftrightarrow CH_2=C(O^-)-OEt$$

This then nucleophilically attacks the carbonyl of another ester molecule:

$$Et-O-C(=O)-CH_2^- + CH_3-C(=O)-OEt \rightleftharpoons Et-O-C(=O)-CH_2-C(CH_3)(O^-)-OEt \rightleftharpoons Et-O-C(=O)-CH_2-C(CH_3)=O + EtO^-$$

$$\xrightarrow{-H^+} CH_3CO-\bar{C}H-COOEt \qquad EtO^- \xrightarrow{+H^+} EtOH$$

> **Connection** Note here the striking similarity to base-promoted ester hydrolysis of section (n). In both cases the leaving group is RO^-, a relatively poor one, which immediately gains a proton at the expense of the other product—in reaction (n), RCOOH gave up its proton to become the resonance-stabilised $RCOO^-$ ion; here, $RCOCH_2COOEt$ gives up a proton to become the resonance-stabilised $[RCOCHCOOEt]^-$ ion [section 2.5]. Again, at the end of the reaction the mixture must be acidified if it is the unionised β-ketoester we actually want as the product.

(q) Reduction of esters

$LiAlH_4$ will reduce an ester in much the same way as it reduces a carboxylic acid.

e.g.

$$H^- \ (AlH_4^-) + CH_3CH_2-C(=O)-OCH_3 \longrightarrow H-C(O^-)(OCH_3)-CH_2CH_3 \longrightarrow H-C(=O)-CH_2CH_3 + CH_3O^-$$

further reduction as in section 5.6(g)

$$\longrightarrow CH_3CH_2CH_2OH$$

(r) Reaction of esters with Grignard reagents

The mechanism here is clearly very similar to the previous one (for reduction).

e.g.

$$CH_3^- \ (CH_3MgBr) + CH_3CH_2-C(=O)-OCH_3 \longrightarrow CH_3-C(O^-)(OCH_3)-CH_2CH_3 \longrightarrow CH_3-C(=O)-CH_2CH_3 + CH_3O^-$$

further reaction as in section 6.5(a)

$$\longrightarrow CH_3-C(CH_3)(OH)-CH_2CH_3$$

(s) Amide hydrolysis under acidic conditions

Amide hydrolysis requires quite severe conditions (e.g. hot 6 M HCl, or boiling NaOH solution). This is because the reactivity of the carbonyl group is much reduced by resonance [see section 10.5(d)]:

$$R-\overset{\delta+}{C}(=O^{\delta-})-\ddot{N}H_2 \longleftrightarrow R-C(O^-)=\overset{+}{N}H_2$$

In concentrated acid the carbonyl group is protonated and rendered more attractive to a nucleophile (or in other words it is rendered more electrophilic).

e.g.

$$H_2O: + CH_3C(NH_2)=\overset{+}{O}H \rightleftharpoons H_2\overset{+}{O}-C(NH_2)(CH_3)-OH \overset{\pm H^+}{\rightleftharpoons} HO-C(\overset{+}{N}H_3)(CH_3)-OH \rightarrow HO-C(CH_3)=\overset{+}{O}H + NH_3$$

$$CH_3CONH_2 \overset{H^+}{\rightleftharpoons} CH_3C(NH_2)=\overset{+}{O}H$$

$$HO-C(CH_3)=\overset{+}{O}H \xrightarrow{-H^+} CH_3COOH \qquad NH_3 \xrightarrow{+H^+} NH_4^+$$

Note how similar this process is to acid-catalysed ester hydrolysis, see section (l). The main difference is that the product now is the basic NH_3 (rather than ROH) and so acid is used up in converting this to NH_4^+ (i.e. amide hydrolysis is not merely *acid-catalysed*).

Exactly the same mechanism is followed in the acid hydrolysis of substituted amides. See if you can write out the detailed mechanism for equations such as:

$$CH_3CONH-C_6H_5 + HCl + H_2O \rightarrow CH_3COOH + C_6H_5-\overset{+}{N}H_3Cl^-$$

$$C_6H_5-CON(CH_3)_2 + HCl + H_2O \rightarrow C_6H_5-COOH + (CH_3)_2\overset{+}{N}H_2Cl^-$$

(t) Amide hydrolysis under basic conditions

Now the nucleophile is OH^- rather than H_2O.

e.g.

$$HO^- + CH_3C(NH_2)=O \rightleftharpoons HO-C(NH_2)(CH_3)-O^- \longrightarrow HO-C(CH_3)=O + NH_2^-$$

$$HO-C(CH_3)=O \xrightarrow{-H^+} CH_3COO^- \qquad NH_2^- \xrightarrow{+H^+} NH_3$$

Note the 'connection' to base-promoted hydrolysis of esters and the Claisen condensation—in each case a poor leaving group immediately gains a proton whilst the other product of the reaction loses one. In the present case, however, NH_2^- is really a very bad leaving group indeed (that is, it is a *very* strong base) and it is more realistic to picture a water molecule assisting the departure of the leaving group so that NH_2^- is never actually produced as such:

$$HO{-}C(NH_2)(CH_3){-}O^- \;\; H{-}OH \longrightarrow HO{-}C(CH_3){=}O + NH_3 + OH^- \xrightarrow{OH^-} CH_3COO^-$$

Again you should be able to follow this mechanism through for hydrolysis of substituted amides such as:

$$CH_3CH_2CONHCH_2CH_3 + NaOH \rightarrow CH_3CH_2COO^-Na^+ + CH_3CH_2NH_2$$

$$C_6H_5{-}CON(CH_3)(CH_2CH_3) + NaOH \rightarrow C_6H_5{-}COO^-Na^+ + CH_3CH_2NHCH_3$$

(u) Imide hydrolysis

This occurs in the Gabriel synthesis of 1° amines [section 5.4(g)]

e.g.

$$\text{phthalimide }(C_6H_4(CO)_2N{-}R) + HO^- \rightleftharpoons \text{tetrahedral intermediate }(C{-}HO,\ O^-) \rightleftharpoons C_6H_4(C(O){-}N^-{-}R)(COOH) \leftrightarrow C_6H_4(C(O^-){=}N{-}R)(COOH) \xrightarrow{\pm H^+} C_6H_4(C(O){-}NHR)(COO^-)$$

In this case the expulsion of the leaving group when the tetrahedral intermediate collapses is relatively easy as the leaving anion is resonance-stabilised as shown. The product above is now of course an amide and further hydrolysis takes place according to the mechanism in section (t) to give:

$$C_6H_4(COO^-\,Na^+)_2 \quad \text{and the 1° amine, } R{-}NH_2$$

(v) Hydrolysis of trihalomethyl ketones

A trihalomethyl ketone is an intermediate in the 'iodoform reaction'.

e.g.

$$CH_3CH_2CH_2{-}\overset{O}{\overset{\|}{C}}{-}CH_3 \xrightarrow{I_2/NaOH} CH_3CH_2CH_2{-}\overset{O}{\overset{\|}{C}}{-}CI_3$$

Subsequently this is nucleophilically attacked by OH^-:

$$HO^- + \underset{R}{\overset{CI_3}{C}}{=}O \rightleftharpoons HO{-}\underset{R}{\overset{CI_3}{C}}{-}O^- \longrightarrow HO{-}\underset{R}{C}{=}O + {}^-CI_3$$

$$HO{-}\underset{R}{C}{=}O \xrightarrow{-H^+} R{-}COO^- \qquad {}^-CI_3 \xrightarrow{+H^+} CHI_3$$

Note: (i) CI_3^- (or CBr_3^- or CCl_3^-) can act as a leaving group whereas CH_3^- cannot [see section 8.4(c)]. That is, a methyl ketone will not be cleaved by alkali until it has been halogenated. (ii) Upon collapse of the tetrahedral intermediate one product then loses a proton and the other gains one—compare with what has already been said above for alkaline hydrolysis of esters and amides, and the Claisen condensation.

(w) Résumé

This has been a long section! To refresh your memory, what you should be appreciating here is that in all these reactions of acyl compounds we have first the attack of a nucleophile on the carbonyl group (or protonated carbonyl group) to give a tetrahedral intermediate, and secondly the expulsion of a leaving group to re-form a (different) acyl compound. Schematically:

$$R{-}\overset{O}{\overset{\|}{C}}{-}X \rightarrow \left[\begin{array}{c}\text{tetrahedral}\\ \text{intermediate}\end{array}\right] \rightarrow R{-}\overset{O}{\overset{\|}{C}}{-}Y$$

(x) Hydrolysis of nitriles

It is convenient to examine the hydrolysis of a nitrile, RCN, in this section because (i) nucleophilic attack on —CN resembles that on a carbonyl group, and (ii) the products of hydrolysis of RCN are the same as those of $RCONH_2$. Water itself does not attack RCN so as above with amides we either first change H_2O to the more nucleophilic OH^- or we change R—CN to the more electrophilic $[R—CNH]^+$.

e.g.

$$HO^- + CH_3C{\equiv}N \longrightarrow HO{-}C(CH_3){=}N^- \xrightarrow{H_2O} HO{-}C(CH_3){=}NH \rightleftharpoons O{=}C(CH_3){-}NH_2$$

or

$$H_2O: + CH_3C{\equiv}\overset{+}{N}H \longrightarrow H_2\overset{+}{O}{-}C(CH_3){=}NH \xrightarrow{-H^+} HO{-}C(CH_3){=}NH \rightleftharpoons O{=}C(CH_3){-}NH_2$$

In both cases the first formed product tautomerises to the more stable amide, $RCONH_2$ (compare with keto–enol tautomerism), which can then be further hydrolysed as in sections (s) or (t). (This makes it obvious of course why hydrolysis of RCN or of $RCONH_2$ leads to the same products.)

5.6 Nucleophilic attack on aldehydes and ketones

We have seen that a nucleophile attacks an **acyl compound** (acid, acyl halide, anhydride, ester, amide, etc.) to give a **tetrahedral intermediate**, which subsequently breaks down with the expulsion of a **leaving group**. With aldehydes and ketones the process is both similar and different—*similar* in that the same kinds of nucleophile attack in the same way to give the same kind of tetrahedral species, but *different* in that now there is no satisfactory leaving group and so we do not get collapse back to a carbonyl-containing compound. (If this were to happen, the leaving group would have to be H^- or R^-, two of the strongest bases and hence the worst leaving groups known.) In brief, with acyl compounds the nucleophile *substitutes*, with aldehydes and ketones the nucleophile *adds*. Schematically:

$$R{-}\overset{O}{\overset{\|}{C}}{-}X \rightarrow \left[\begin{array}{c}\text{tetrahedral}\\ \text{intermediate}\end{array}\right] \rightarrow R{-}\overset{O}{\overset{\|}{C}}{-}Y$$

$$\left.\begin{array}{c} R{-}\overset{O}{\overset{\|}{C}}{-}H \\ R{-}\overset{O}{\overset{\|}{C}}{-}R' \end{array}\right\} \rightarrow \begin{array}{c}\text{tetrahedral}\\ \text{species}\end{array} \quad ---\!\!\times\!\!--\rightarrow$$

We will illustrate this with several examples; note how similar the mechanism is in all cases.

(a) Addition of Grignard reagents [see also sections 2.7(b) and 6.5(a)]

e.g.

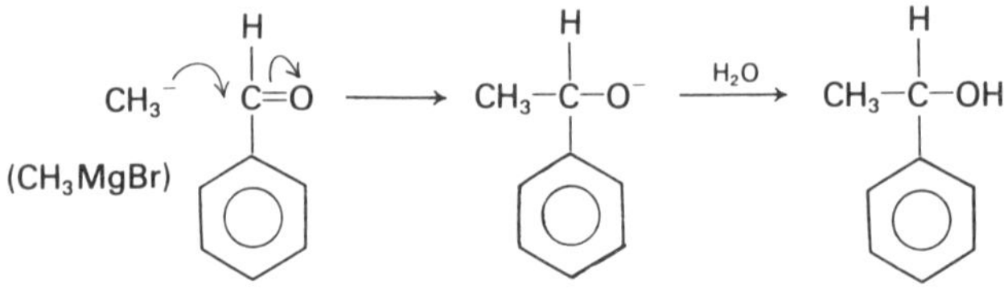

$$CH_3^- \; (CH_3MgBr) + C_6H_5-CH{=}O \longrightarrow CH_3-CH(C_6H_5)-O^- \xrightarrow{H_2O} CH_3-CH(C_6H_5)-OH$$

(b) The Reformatski reaction [see also section 2.11, 10.3(c)]

e.g.

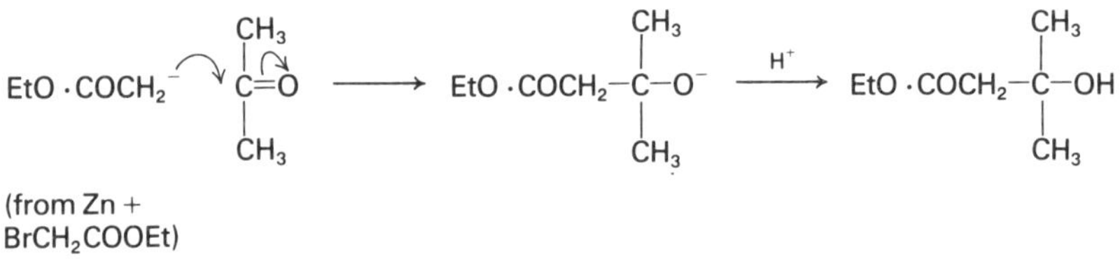

$$EtO\cdot COCH_2^- + (CH_3)_2C{=}O \longrightarrow EtO\cdot COCH_2-C(CH_3)_2-O^- \xrightarrow{H^+} EtO\cdot COCH_2-C(CH_3)_2-OH$$

(from Zn + $BrCH_2COOEt$)

(c) Addition of alkynide ions

e.g.

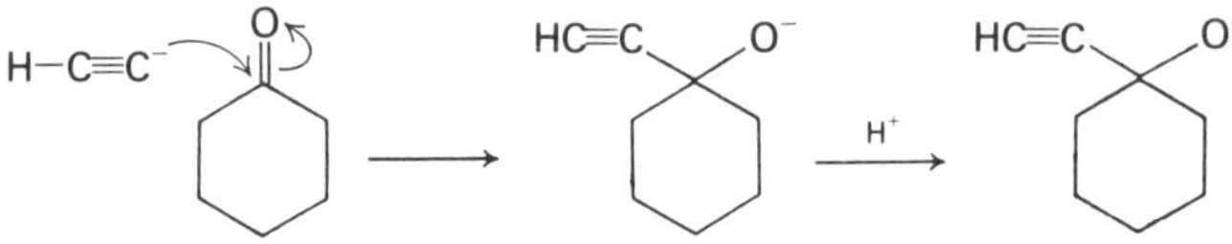

$$H-C{\equiv}C^- + \text{cyclohexanone} \longrightarrow \text{1-(HC}{\equiv}\text{C)-cyclohexyl-}O^- \xrightarrow{H^+} \text{1-(HC}{\equiv}\text{C)-cyclohexyl-}OH$$

(d) Addition of hydrogen cyanide

e.g.

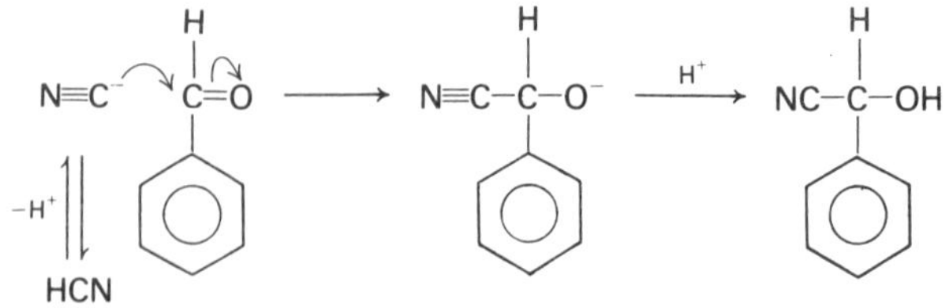

$$HCN \underset{}{\overset{-H^+}{\rightleftharpoons}} N{\equiv}C^- ; \quad N{\equiv}C^- + C_6H_5-CH{=}O \longrightarrow N{\equiv}C-CH(C_6H_5)-O^- \xrightarrow{H^+} NC-CH(C_6H_5)-OH$$

(a cyanohydrin)

(e) The aldol condensation [see also section 2.11, 6.5(h)]

e.g.

$$CH_3CH_2-\overset{O}{\overset{\|}{C}}-H \underset{}{\overset{OH^-}{\rightleftharpoons}} CH_3\bar{C}H-\overset{O}{\overset{\|}{C}}-H \longleftrightarrow CH_3CH=\overset{O^-}{\overset{|}{C}}-H$$

Then:

$$\text{(H-C(=O))(CH}_3\text{)CH}^- + \text{H-C(=O)-CH}_2\text{CH}_3 \rightleftharpoons \text{(H-C(=O))(CH}_3\text{)CH-C(H)(CH}_2\text{CH}_3\text{)-O}^- \xrightarrow{H_2O} \text{(H-C(=O))(CH}_3\text{)CH-C(H)(CH}_2\text{CH}_3\text{)-OH}$$

$$(\text{or } CH_3CH_2\underset{OH}{\underset{|}{C}H}-\underset{CH_3}{\underset{|}{C}H}CHO)$$

Notice how very similar this reaction is in its early stages to the Claisen condensation [section 5.5(p)]. In both cases we have first the generation of a resonance-stabilised carbanion. This then attacks (as a nucleophile) the carbonyl group of an unionised molecule of the original compound to give a tetrahedral species. For the aldol case, however, this is the end of the reaction (apart from picking up a proton from the solvent), whereas in the Claisen condensation there is a suitable leaving group (EtO^-) and so the tetrahedral intermediate collapses to re-form a carbonyl group.

(f) The Wittig reaction

[See also sections 2.11 and 6.6(p).]

e.g.

$$\underset{^+P(Ph)_3}{\underset{|}{CH_2^-}} + \overset{CH_3}{\overset{|}{C}}\underset{CH_2CH_3}{\underset{|}{=O}} \longrightarrow Ph_3\overset{+}{P}-CH_2-\overset{CH_3}{\overset{|}{\underset{CH_2CH_3}{\underset{|}{C}}}}-O^-$$

$$\updownarrow$$

$$CH_2=P(Ph)_3$$

Clearly the first step is the same as all those in this section—nucleophilic attack on an aldehyde or ketone to give a tetrahedral intermediate. Then this particular reaction takes off on a course of its own:

$$\underset{Ph_3P^+}{\underset{|}{CH_2}}-\overset{CH_3}{\overset{|}{\underset{^-O}{\underset{|}{C}}}}-CH_2CH_3 \longrightarrow \underset{Ph_3P}{\underset{|}{CH_2}}-\overset{CH_3}{\overset{|}{\underset{O}{\underset{|}{C}}}}-CH_2CH_3 \longrightarrow CH_2=\overset{CH_3}{\overset{|}{C}}-CH_2CH_3 + Ph_3P=O$$

(g) Reduction of aldehydes and ketones

$LiAlH_4$ and $NaBH_4$ can both be thought of as furnishing H^-.

e.g.

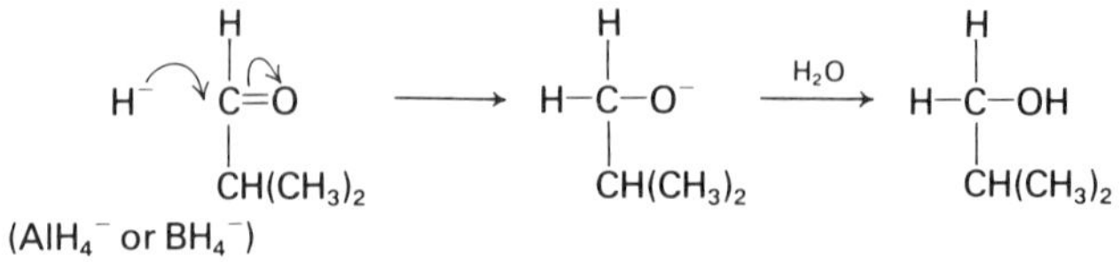

$$H^- + H\text{-}C(=O)\text{-}CH(CH_3)_2 \longrightarrow H\text{-}CH(O^-)\text{-}CH(CH_3)_2 \xrightarrow{H_2O} H\text{-}CH(OH)\text{-}CH(CH_3)_2$$

(AlH_4^- or BH_4^-)

Remember H^- is not actually free in this reaction; it is transferred to the carbonyl group from AlH_4^- or BH_4^- in the same way as R^- is transferred from RMgX in a Grignard reaction.

(h) Addition of water

When an aldehyde or ketone dissolves in water, a nucleophilic addition of water takes place to set up an equilibrium between the original carbonyl compound and its hydrate.

e.g.

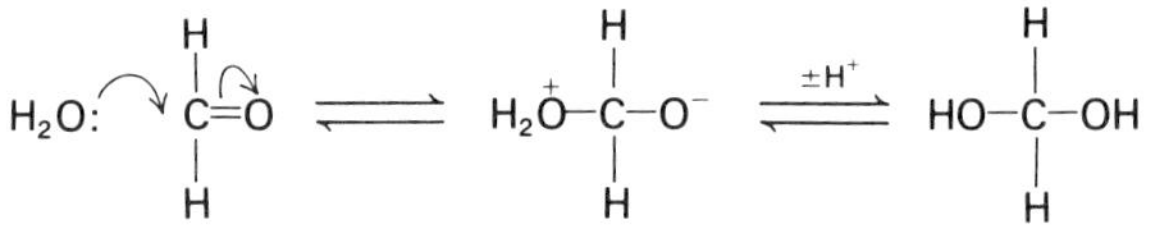

$$H_2O: + H_2C{=}O \rightleftharpoons H_2\overset{+}{O}\text{-}CH_2\text{-}O^- \overset{\pm H^+}{\rightleftharpoons} HO\text{-}CH_2\text{-}OH$$

With formaldehyde (methanal) as shown here, the position of the equilibrium lies well over to the right, with benzaldehyde it lies well to the left, and with acetaldehyde (ethanal) there is some intermediate percentage (~20–30) of hydrate present. In all cases with one or two exceptions [e.g. see section 9.5(c)] the hydration is freely reversible and the hydrate cannot be isolated.

(i) Addition of alcohols

When an aldehyde or ketone dissolves in an alcohol, a nucleophilic addition takes place to some extent (just as in the previous case with water) to give a hemiacetal or hemiketal.

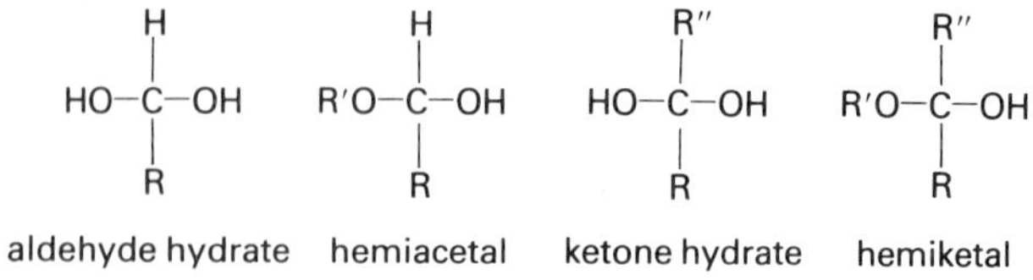

HO–CH(R)–OH	R′O–CH(R)–OH	HO–C(R″)(R)–OH	R′O–C(R″)(R)–OH
aldehyde hydrate	hemiacetal	ketone hydrate	hemiketal

Again these products generally cannot be isolated. However, if an aldehyde is mixed with an alcohol in the presence of dry HCl, the reaction goes further to give an acetal (which *is* stable). The mechanism is:

$$R'\ddot{O}H + HC(R)=\overset{+}{O}H \rightleftharpoons R'-\overset{+}{O}(H)-CH(R)-OH \overset{\pm H^+}{\rightleftharpoons} R'-\ddot{O}-CH(R)-\overset{+}{O}H_2 \overset{-H_2O}{\rightleftharpoons} R'-\overset{+}{O}=CH(R)$$

$$RCHO \overset{H^+}{\rightleftharpoons} HC(R)=\overset{+}{O}H$$

Then:

$$R'-\ddot{O}H + HC(R)=\overset{+}{O}-R' \rightleftharpoons R'-\overset{+}{O}(H)-CH(R)-OR' \overset{-H^+}{\rightleftharpoons} R'O-CH(R)-OR'$$

Note: (i) The two key steps of the process are very similar, namely nucleophilic attack of R′OH on either

$$\text{H(R)C}=\overset{+}{O}H \quad \text{or} \quad \text{H(R)C}=\overset{+}{O}R'$$

to give tetrahedral species.

(ii) As shown, these transformations are all equilibria. The reaction can be driven to give a good yield of acetal by using an excess of R′OH. Conversely, the acetal can be hydrolysed back to aldehyde and alcohol by dilute acid.

(iii) Theoretically a ketone will react in exactly the same way. However, in this case the equilibrium does not lie so well over to the product side, probably for simple steric reasons:

$$R'O-C(R'')(R)-OR' \text{ has more steric crowding than } R'O-C(H)(R)-OR'$$

A good yield of ketal *can*, however, be achieved with a diol (or dithiol) since there is a very favourable entropy term to the energetics of the reaction when both —OH (or —SH) groups are furnished by the same molecule. The product in cases such as these is a 5- or 6-membered cyclic ketal (or thioketal) such as:

$$R_2C(O-CH_2)_2CH_2 \quad \text{and} \quad R_2C(S-CH_2)_2$$

(Can you follow all steps in the mechanism, like that shown above, when the starting materials are, say, cyclohexanone and 1,2-ethanedithiol?)

(iv) Again for reasons of favourable entropy a **cyclic** hemiacetal or hemiketal is more stable than an acyclic one. Common examples of these of course occur in carbohydrate chemistry.

e.g.

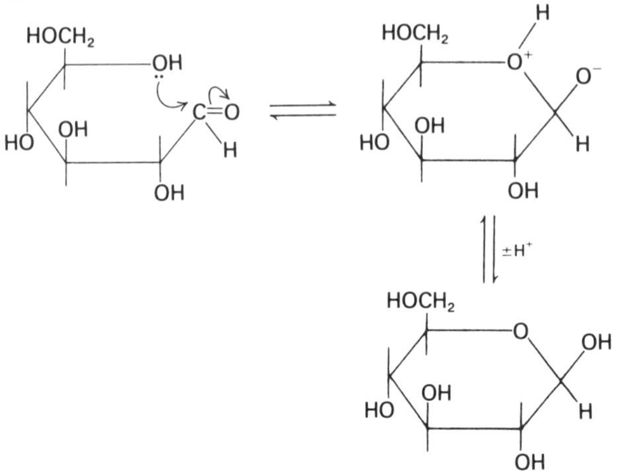

(v) I have said above that after a nucleophile has attacked an aldehyde or ketone the tetrahedral species does not possess a leaving group. It is certainly true that R^- or H^- does not act as a leaving group, but strictly speaking we should not neglect the other two substituents round the tetrahedral carbon atom. One of these is the incoming nucleophile and that *can* often act as a leaving group, a process that merely leads of course back to the starting materials; for example, consider the reverse of the first step of acetal formation:

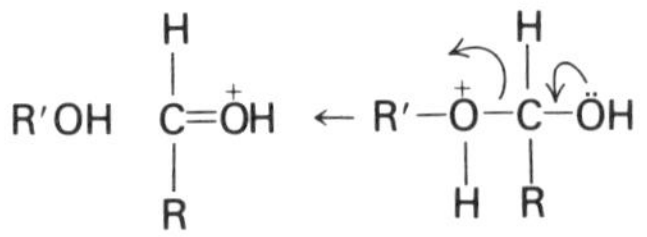

$$R'OH \quad HRC{=}\overset{+}{O}H \leftarrow R'{-}\overset{+}{O}(H){-}CHR{-}\ddot{O}H$$

The *fourth* substituent in the tetrahedral intermediate is always $-O^-$ or $-OH$ and if this can be converted to $-OH_2^+$ then we have another potential leaving group. It is precisely this process that happens in acetal formation; the protonated hemiacetal loses a molecule of water, and the resulting oxonium ion is attacked by a second molecule of R′OH.

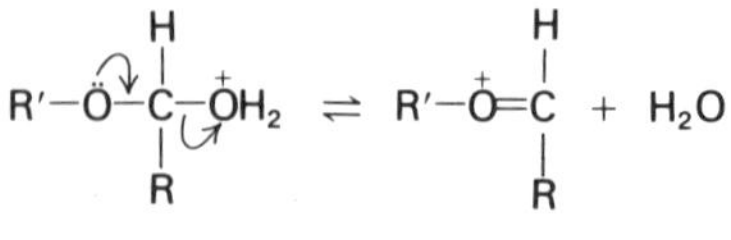

$$R'{-}\ddot{O}{-}CHR{-}\overset{+}{O}H_2 \rightleftharpoons R'{-}\overset{+}{O}{=}CHR + H_2O$$

(j) Addition of ammonia and derivatives of ammonia

e.g.

$$H_3\ddot{N} + HRC{=}O \rightleftharpoons H_3\overset{+}{N}{-}CHR{-}O^- \overset{\pm H^+}{\rightleftharpoons} H_2N{-}CHR{-}OH$$

an 'aldehyde–ammonia' (compare with an aldehyde hydrate)

Then:

$$H_2N{-}CHR{-}OH \overset{H^+}{\rightleftharpoons} H_2\ddot{N}{-}CHR{-}\overset{+}{O}H_2 \overset{-H_2O}{\rightleftharpoons} H_2\overset{+}{N}{=}CHR \overset{-H^+}{\rightleftharpoons} HN{=}C(H)R$$

an imine

[Compare:

$$R'O{-}CHR{-}OH \overset{H^+}{\rightleftharpoons} R'{-}\ddot{O}{-}CHR{-}\overset{+}{O}H_2 \overset{-H_2O}{\rightleftharpoons} R'{-}\overset{+}{O}{=}CHR$$
]

I have repeated in brackets the appropriate part of the mechanism involved in the conversion of a hemiacetal to an acetal, to stress the 'connection'. In both cases, a molecule of H_2O acts as the leaving group.

Generally speaking neither the tetrahedral product nor the imine is formed in good yield with ammonia itself, but certain derivatives of ammonia do lead to high yields of the appropriate imine. See Table 5.1.

Table 5.1

Reagent, $X{-}NH_2$	Product, $X{-}N{=}C<$
$HO{-}NH_2$	oxime
$H_2N{-}NH_2$	hydrazone
$PhHN{-}NH_2$	phenylhydrazone
$2,4\text{-}(NO_2)_2PhHN{-}NH_2$	2,4-dinitrophenylhydrazone
$H_2N{-}CO{-}HN{-}NH_2$	semicarbazone

The rate of reaction between an aldehyde or ketone and these reagents is pH-sensitive. With decreasing pH the rate increases because the dehydration step is acid-catalysed as shown above (and this is the rate-determining step). However, if the pH is too low the nucleophile is protonated (destroying its nucleophilicity) and the rate of reaction falls:

$$X{-}\ddot{N}H_2 + H^+ \rightleftharpoons X{-}\overset{+}{N}H_3$$

(is a nucleophile) (not a nucleophile; can no longer furnish a pair of electrons)

(k) The Cannizzaro reaction

This occurs with aldehydes lacking α-hydrogens when treated with concentrated hydroxide solution:

e.g.

$$2\,C_6H_5\text{—CHO} \xrightarrow{OH^-} C_6H_5\text{—COO}^- + C_6H_5\text{—CH}_2OH$$

The mechanism involves as a first step the nucleophilic attack of OH^-:

$$HO^- + \underset{R}{\overset{H}{|}}C{=}O \rightleftharpoons HO\text{—}\underset{R}{\overset{H}{|}}C\text{—}O^-$$

The resulting anion then transfers a hydride ion to another molecule of the aldehyde—a second nucleophilic attack giving a second tetrahedral intermediate (the alkoxide ion, RCH_2O^-):

$$^-O\text{—}\underset{R}{\overset{OH}{|}}C\text{—}H + \underset{R}{\overset{H}{|}}C{=}O \longrightarrow O{=}\underset{R}{\overset{OH}{|}}C + H\text{—}\underset{R}{\overset{H}{|}}C\text{—}O^-$$

$$\downarrow -H^+ \qquad \downarrow +H^+$$

$$RCOO^- \qquad RCH_2OH$$

This is an unusual reaction in that the first formed tetrahedral intermediate collapses with the expulsion of H^- as the leaving group (and as a general rule we have stated that H^- is too poor a leaving group to take on this role). However, H^- is not expelled as such, but is transferred directly to the other aldehyde molecule, in just the same way as H^- can be transferred to RCHO from BH_4^- or AlH_4^- [section 5.6(g)]. We can think of the electrophilic carbonyl group of the second aldehyde molecule assisting the departure of H^- in much the same way as we have pictured a water molecule assisting the departure of NH_2^- in alkaline hydrolysis of amides [section 5.5(t)].

(l) Résumé

Nucleophilic attack on an aldehyde or ketone produces a tetrahedral species which does not expel a leaving group. Exceptions to this general statement are:

- (i) the Cannizzaro reaction
- (ii) cases where a molecule of water can be lost, as in the formation of imines
- (iii) the haloform reaction, where the leaving group is CX_3^-.

5.7 Synopsis

The nucleophilic attack of A on B represents a very large fraction of all the reactions in organic chemistry. It is difficult to summarise such a large subject briefly but an attempt at this has been made in Table 5.2. It is designed to jog your memory of the major classes of nucleophilic attack. There is no room obviously for much detail; if an entry fails to remind you of the appropriate details then you should look back through the preceding pages and revise them.

Table 5.2

	Result of nucleophilic attack on:		
Nucleophile	Alkyl halides	Aldehydes/ ketones	Acyl compounds
H_2O or OH^-	alcohols	hydrates	hydrolysis (of $RCOCl$, $RCONH_2$, $RCOOR'$, etc.)
ROH or RO^-	ethers (Williamson)	acetals/ketals	esters
NH_3, $R{-}NH_2$, etc.	amines	*2,4-d.n.p., oximes, etc.	amides
CN^- or HCN	nitriles	cyanohydrins	—
organometallic reagents	alkanes (Corey–House)	alcohols (1°, 2° and 3°)	3° alcohol from RMgX and ester; ketone from R_2Cd and acyl halide
$RCOO^-$	ester	—	anhydride (from acyl halide)
$RC{\equiv}C^-$	$R{-}C{\equiv}C{-}R'$	addition product	—
various enolate anions	e.g. $CH_3COCH(R'){-}COOEt$	aldols; mixed aldols	β-ketoesters (Claisen condensation)
$LiAlH_4$	alkanes	1° and 2° alcohols	1° alcohols (from acids and esters)

*2,4-d.n.p. = 2,4-diphenylhydrazone

5.8 Nucleophilic attack on α,β-unsaturated carbonyl compounds

[See also section 2.12.] With α,β-unsaturated carbonyl compounds a nucleophile may attack at the carbonyl group in the usual way as we have seen in sections 5.5 and 5.6, or alternatively it may attack at the β-position:

Either:

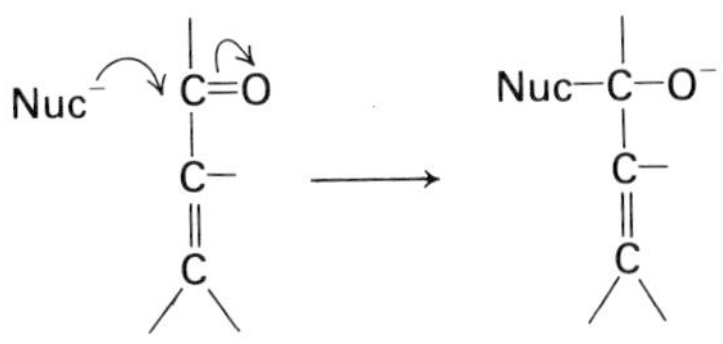

$$Nuc^- + >C{=}C{-}C{=}O \longrightarrow >C{=}C{-}C(Nuc){-}O^-$$

Or:

$$\mathrm{Nuc^- + C{=}C{-}C{=}O \longrightarrow Nuc{-}C{-}C{=}C{-}O^-}$$

If it follows the latter course the subsequent steps in the reaction are protonation to give the enol followed by rearrangement (tautomerisation) to the keto form:

$$\mathrm{Nuc{-}C{-}C{=}C{-}O^- \xrightarrow{H^+} Nuc{-}C{-}C{=}C{-}OH \rightleftharpoons Nuc{-}C{-}C(H){-}C{=}O}$$

Whether the nucleophile attacks the carbonyl or the β-carbon depends on the reactants; sometimes one product predominates, sometimes the other.

e.g.

$$\mathrm{CH_3CH{=}CHCHO + CH_2(COOH)_2 \xrightarrow{pyridine} CH_3CH{=}CH{-}C(OH)(H){-}CH(COOH)_2}$$

Here we have attack on the **carbonyl group**. [What is the nucleophile? See section 10.3(e).] The product then loses H_2O and CO_2 to give finally sorbic acid (hexa-2,4-dienoic acid, $CH_3CH{=}CH{-}CH{=}CH{-}COOH$). On the other hand, with the following similar starting materials we get attack mainly at the ***β*-position**:

$$\mathrm{PhCH{=}CH{-}C({=}O){-}Ph + CH_2(COOEt)_2 \xrightarrow{piperidine} Ph{-}CH(CH(COOEt)_2){-}CH{-}C({=}O){-}Ph}$$

Consider also the addition of HCN to 2-butenal:

$$\mathrm{CH_3CH{=}CH{-}C({=}O){-}H + HCN \rightarrow CH_3CH{=}CH{-}C(OH)(CN){-}H}$$

and to butenone:

$$\mathrm{CH_3{-}C({=}O){-}CH{=}CH_2 + HCN \rightarrow CH_3{-}C({=}O){-}CH_2{-}CH_2CN}$$

In these examples, attack at the carbonyl group is favoured with α,β-unsaturated aldehydes, whereas attack at the β-carbon occurs with α,β-unsaturated ketones. For a ketone, the observed product may be preferred for steric reasons:

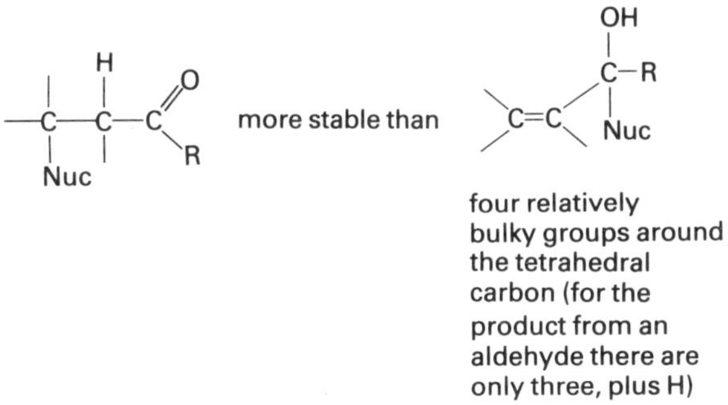

more stable than

four relatively bulky groups around the tetrahedral carbon (for the product from an aldehyde there are only three, plus H)

Moreover, the polarity of the carbonyl group in a ketone is less than in an aldehyde because of the inductive effect of the alkyl group [section 9.3] and this may tend to inhibit nucleophilic attack at this position:

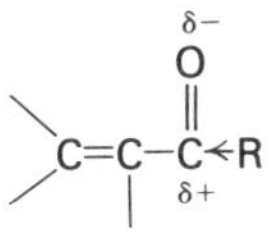

5.9 Nucleophilic aromatic substitution

An **aryl** halide is much less reactive than an **alkyl** halide. The orbitals containing the halogen atom's lone pairs of electrons can overlap to some extent with the delocalised π-electron orbitals of the benzene ring. The result is that the C–halogen bond in an aryl halide is shorter and stronger than in an alkyl halide. Consequently, a compound such as bromobenzene will not undergo reactions involving either the S_N1 or S_N2 mechanisms. For example, the following processes (a) and (b) do **not** happen:

(a) $C_6H_5Br \not\rightleftharpoons C_6H_5^+ + Br^-$; $C_6H_5^+ \xrightarrow{H_2O} C_6H_5OH$; $Br^- \xrightarrow{Ag^+} AgBr$

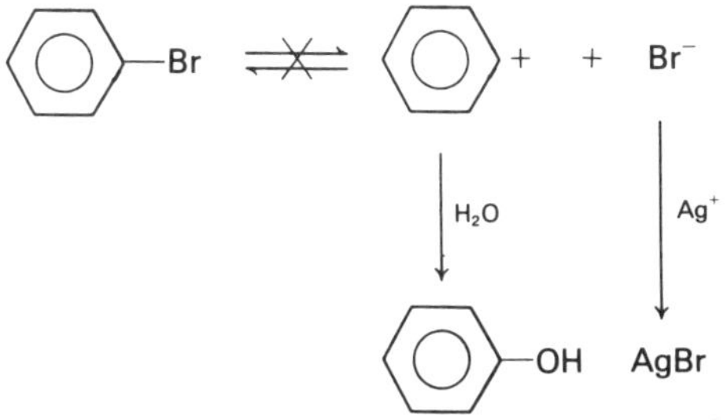

unlike, say, $(CH_3)_3C—Br$ or $PhCH_2Br$ which *do* show S_N1 reactivity [section 1.9]

(b)

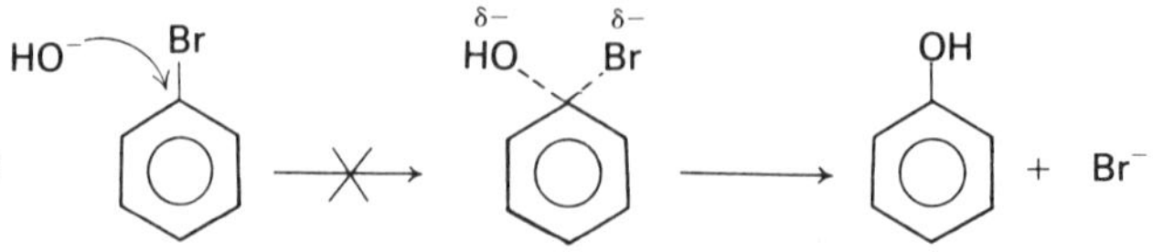

unlike, say, CH_3CH_2Br which *does* show S_N2 reactivity.

Some aryl halides, however, are reactive. The classic example is picryl chloride (2,4,6-trinitrochlorobenzene) which is readily hydrolysed by water [see also section 10.8(b)]:

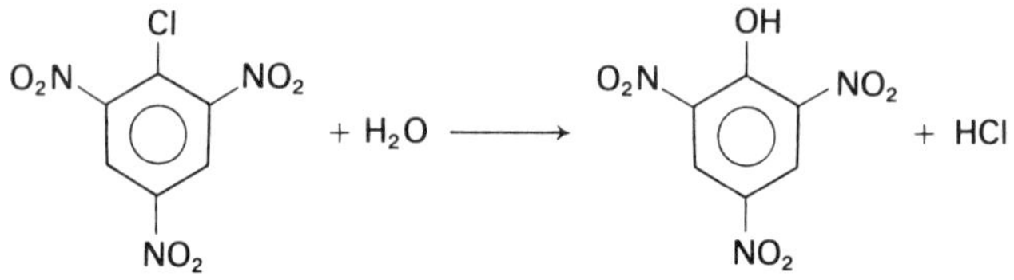

The mechanism is still neither S_N1 nor S_N2. In fact it involves nucleophilic attack to give an intermediate that subsequently expels a leaving group (Cl^-):

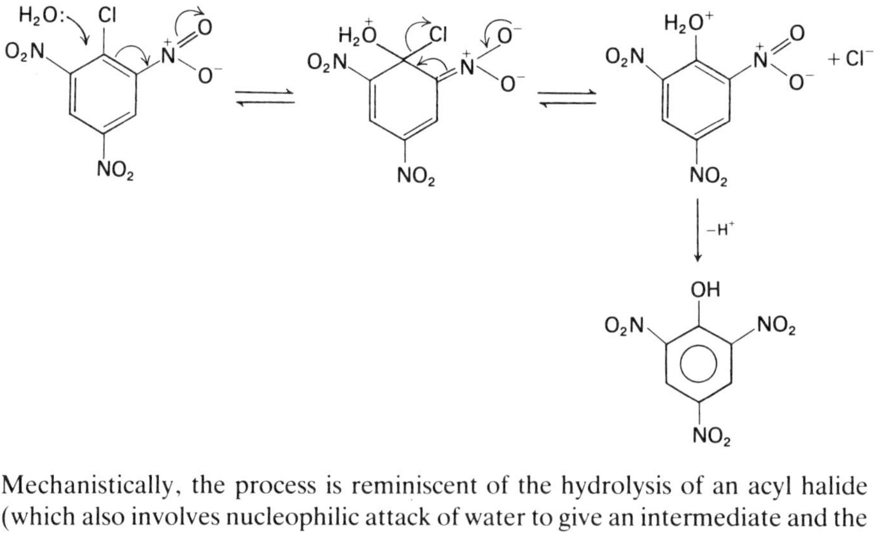

Mechanistically, the process is reminiscent of the hydrolysis of an acyl halide (which also involves nucleophilic attack of water to give an intermediate and the subsequent expulsion of Cl^-):

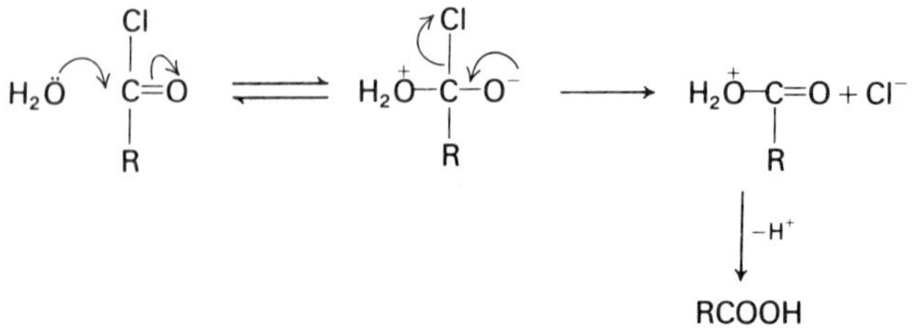

I have drawn above only one of the resonance forms for the intermediate involved when water attacks picryl chloride; the negative charge could equally well be pictured taken up by the other $-NO_2$ groups as well. (Can you draw the resulting resonance forms?)

If one of the three nitro groups is missing the compound is less readily hydrolysed, and if two are missing the reaction conditions need to be rather severe:

$$p\text{-}ClC_6H_4NO_2 \xrightarrow[160\,°C]{15\%\ NaOH} p\text{-}HOC_6H_4NO_2$$

(Chlorobenzene itself can only be hydrolysed at very high temperature and pressure.)

5.10 Nitrosation of amines

We have considered the nucleophilic attack of amines on carbonyl compounds, [sections 5.5(d), (e), 5.6(j)]. Very similar (initially, at least) is the nucleophilic attack of amines on N_2O_3 or $ONOH_2^+$ (which are the active species in nitrosation):

$$NaNO_2 + HCl \rightarrow NaCl + HO{-}N{=}O \quad \text{(nitrous acid)}$$

$$HO{-}N{=}O + H^+ \rightarrow H_2\overset{+}{O}{-}N{=}O$$

$$2HO{-}N{=}O \rightleftharpoons O{=}N{-}O{-}N{=}O + H_2O$$

Then:

$$R{-}\ddot{N}H_2 + \overset{+}{O}H_2{-}N{=}O \rightleftharpoons R{-}\overset{+}{N}H_2{-}N(\overset{+}{O}H_2){-}O^- \longrightarrow R{-}\overset{+}{N}H_2{-}N{=}O + H_2O$$

Or:

$$R{-}\ddot{N}H_2 + ONO{-}N{=}O \rightleftharpoons R{-}\overset{+}{N}H_2{-}N(ONO){-}O^- \longrightarrow R{-}\overset{+}{N}H_2{-}N{=}O + NO_2^-$$

Compare:

$$\left[R{-}\ddot{N}H_2 + Cl{-}C(R){=}O \rightleftharpoons R{-}\overset{+}{N}H_2{-}C(Cl)(R){-}O^- \longrightarrow R{-}\overset{+}{N}H_2{-}C(R){=}O + Cl^-\right]$$

Then:

$$R\overset{+}{N}H_2{-}N{=}O \xrightarrow{-H^+} R{-}NH{-}N{=}O \overset{\text{tautomerisation}}{\rightleftharpoons} R{-}N{=}N{-}OH$$

$$R{-}N{=}N{-}OH \xrightarrow{H^+} R{-}\ddot{N}{=}N{-}\overset{+}{O}H_2 \xrightarrow{-H_2O} R{-}\overset{+}{N}{\equiv}N$$

a diazonium ion

This is what happens with a 1° amine; the product is only stable if R is aromatic [sections 1.2, 10.9]. With a 2° amine, the reaction proceeds in a similar way as far as R(R′)N—N=O which separates as a yellow oil or solid. (This compound has no H on the nitrogen so it cannot tautomerise and react further as in the primary case.) With a 3° aliphatic amine the reaction is complex. With a 3° aromatic amine the product is:

$$R_2N{-}C_6H_4{-}N{=}O$$

This is formed by electrophilic substitution into the activated benzene ring [compare section 4.5].

6

Carbon–Carbon Bond Formation

6.1 Introduction

In the preceding chapters we have already met most of the ways of synthesising new C—C bonds. Hence this chapter is an opportunity to gather the various methods together and classify them according to the type of chemical process they involve.

In the main there are only two general ways of making a C—C bond which I will symbolise as:

$$(1) \qquad C\cdot \quad + \quad C\cdot \quad \rightarrow \quad C{-}C$$
$$(2) \qquad C(+) \quad + \quad C(-) \quad \rightarrow \quad C{-}C$$

Process (1) involves the dimerisation of free radicals and other free radical reactions, whereas in process (2) the symbols C(+) and C(−) *do not necessarily* mean species with a full positive or negative charge, but rather represent an electrophile and a nucleophile respectively. Thus C(+) might be for instance $R^{\delta+}{-}Cl$ or $R_2C^{\delta+}{=}O$, and C(−) might be $R^{\delta-}MgBr$.

Process (2) is much more common and important than process (1). The vast majority of C—C bond-synthesising reactions are essentially the combination of an electrophile and a nucleophile. Electrophilic carbon is found in carbocations, alkyl halides and carbonyl compounds and we will consider nucleophilic attack on each of these categories, but first we will look at free radical C—C bond formation.

6.2 Free radical C—C bond formation

(a) Dimerisations

(i) The simplest conceivable C—C bond-forming reaction is that which occasionally happens as a **termination** step in the free radical halogenation of methane [section 3.2]:

$$CH_3\cdot + CH_3\cdot \rightarrow CH_3{-}CH_3$$

The main products of this reaction are of course $CH_3{-}X$ and H—X. Since the concentration of $CH_3\cdot$ at any given time is very low it is only infrequently that two of them collide; hence such a process does not constitute a practical synthesis of ethane.

(ii) An adaption of this free radical dimerisation reaction that does constitute a practical synthesis is the **Kolbé** electrolytic method [section 3.11]:

$$\begin{aligned} R{-}COO^- &\rightarrow R{-}COO\cdot + e^- \\ R{-}COO\cdot &\rightarrow R\cdot + CO_2 \\ 2R\cdot &\rightarrow R{-}R \end{aligned}$$

(iii) Next, consider the reduction of acetone (propanone) to pinacol (what is its systematic name?) by amalgamated magnesium:

$$2CH_3COCH_3 \xrightarrow{Mg(Hg)} CH_3{-}C(CH_3)(OH){-}C(CH_3)(OH){-}CH_3$$

The first step here is that a magnesium atom donates an electron to each of two molecules of acetone:

$$2\,CH_3{-}CO{-}CH_3 + :Mg \longrightarrow 2\,CH_3{-}\dot{C}(O^-){-}CH_3 + Mg^{++}$$

The resulting species is both a **radical** and an **anion**. (Convince yourself that you can account for all the electrons and the negative charge.) The two radical ions then dimerise:

$$CH_3{-}\dot{C}(CH_3){-}O^- + \cdot C(CH_3)(O^-){-}CH_3 \longrightarrow CH_3{-}C(CH_3)(O^-){-}C(CH_3)(O^-){-}CH_3 \xrightarrow{H_2O} \text{pinacol}$$

Note the great tendency for two unpaired electrons to pair up and form a bond even at the expense of bringing together two negatively charged species.

(iv) A similar mechanism is found in the acyloin condensation. This starts with an ester rather than a ketone and sodium in place of magnesium:

$$R{-}C(=O){-}OCH_3 + \cdot Na \longrightarrow R{-}\dot{C}(O^-){-}OCH_3 + Na^+$$

Again, the two radical ions then dimerise:

$$R-\overset{\overset{OCH_3}{|}}{\underset{\underset{O^-}{|}}{C}}\cdot \; + \; \cdot\overset{\overset{OCH_3}{|}}{\underset{\underset{O^-}{|}}{C}}-R \longrightarrow R-\overset{\overset{OMe}{|}}{\underset{\underset{O^-}{|}}{C}}-\overset{\overset{OMe}{|}}{\underset{\underset{O^-}{|}}{C}}-R$$

Now however, unlike the previous reaction, we have a leaving group present (MeO^-). There is nothing surprising about this—we have seen many reactions of esters and of ketones in which there were and were not leaving groups respectively. Hence the next step is:

$$R-\overset{\overset{OMe}{|}}{\underset{\underset{O^-}{|}}{C}}-\overset{\overset{OMe}{|}}{\underset{\underset{O^-}{|}}{C}}-R \longrightarrow R-\underset{\underset{O}{\|}}{C}-\underset{\underset{O}{\|}}{C}-R + 2MeO^-$$

And then the diketone is reduced to an enediol that tautomerises to the product **acyloin** (or α-hydroxyketone):

$$R-\underset{\underset{O}{\|}}{C}-\underset{\underset{O}{\|}}{C}-R \longrightarrow R-\underset{\underset{OH}{|}}{C}=\underset{\underset{OH}{|}}{C}-R \rightleftharpoons R-\underset{\underset{O}{\|}}{C}-\overset{\overset{H}{|}}{\underset{\underset{OH}{|}}{C}}-R$$

This reaction has been of some use in synthesising large rings from long chain diesters:

$$(CH_2)_n(COOR)_2 \xrightarrow{Na} (CH_2)_n \text{ ring closed by } C{=}O \text{ and } CH{-}OH$$

If n were only 4 or 5, one could do one of the common types of cyclisation reaction here [such as the Dieckmann reaction; see section 6.5(m)], but if n is greater than this, the entropy of reaction becomes unfavourable. In simple terms this means the chance of the two 'reactive ends' of the molecule ever coming close enough to react is too small for the reaction to proceed at a reasonable rate. However, in the acyloin condensation both ends of the molecule are adsorbed on the surface of the sodium metal enhancing their chances of reacting with each other.

(v) Terminal alkynes can also be made to dimerise (in a process that probably involves free radicals). The reaction is known as **oxidative coupling** and requires a copper compound.
e.g.

$$2R-C\equiv C-H \xrightarrow[\text{pyridine}]{CuCl,\ O_2} R-C\equiv C-C\equiv C-R$$

This reaction too has been adapted for synthesising cyclic compounds, notably **annulenes**:

e.g.

$$3\ HC{\equiv}C{-}CH_2{-}CH_2{-}C{\equiv}CH \xrightarrow[\text{coupling}]{\text{oxidative}} \text{cyclic trimer } (CH_2CH_2{-}C{\equiv}C{-}C{\equiv}C{-})_3 \xrightarrow[\text{steps}]{\text{several}} \text{[18]-annulene}$$

(+ tetramer, pentamer, etc.)

[18]-annulene

(b) Free radical polymerisation of alkenes

[See also section 3.10.] The essential factor here, like the dimerisations we have just been looking at, is that each of the two reactants supplies one electron for the new C—C bond. For example, in the polymerisation of styrene:

$$\text{---}CH_2{-}\underset{Ph}{CH}{-}CH_2{-}\underset{Ph}{CH}{-}CH_2{-}\underset{Ph}{CH}\cdot \quad CH_2{=}\underset{Ph}{CH} \rightarrow \text{---}CH_2{-}\underset{Ph}{CH}{-}CH_2{-}\underset{Ph}{CH}{-}CH_2{-}\underset{Ph}{CH}{-}CH_2{-}\underset{Ph}{CH}\cdot$$

(Industrially, such C—C bond-forming reactions are extremely important.)

6.3 Nucleophilic attack on carbocations

We are now entering that large group of C—C bond-forming reactions in which one reactant (the nucleophile) provides both of the electrons for the new bond.

(a) Acid-catalysed polymerisation of alkenes

[See also section 1.7(d).]

e.g.

$$\text{---}CH_2{-}\underset{CH_3}{CH}{-}CH_2{-}\underset{CH_3}{CH}{-}CH_2{-}\underset{CH_3}{\overset{+}{C}H} \quad CH_2{=}\underset{CH_3}{CH} \rightarrow \text{---}CH_2{-}\underset{CH_3}{CH}{-}CH_2{-}\underset{CH_3}{CH}{-}CH_2{-}\underset{CH_3}{CH}{-}CH_2{-}\underset{CH_3}{\overset{+}{C}H}$$

Here the alkene is the nucleophile, and the carbocation at the end of the growing polymer chain is the electrophile.

Under the right conditions the reaction can be limited to dimerisation rather than polymerisation.

e.g.

$$CH_3-C(=CH_2)-CH_3 \xrightarrow{H^+} CH_3-C^+(CH_3)-CH_3 \quad CH_2=C(CH_3)_2 \longrightarrow CH_3-C(CH_3)_2-CH_2-C^+(CH_3)_2$$

$$\xrightarrow{-H^+} (CH_3)_3C-CH=C(CH_3)_2 \text{ (major)} \quad + \quad (CH_3)_3C-CH_2-C(=CH_2)CH_3 \text{ (minor)}$$

(Note: The more alkyl groups carried by the double bond, the more stable is the alkene, provided the alkyl groups are not too bulky.)

(b) Friedel–Crafts alkylation and acylation

e.g.

$$C_6H_6 \text{ (nucleophile)} + \overset{\delta+}{CH_3}{-}{-}{-}{-}\overset{\delta-}{AlCl_4} \text{ (electrophile)} \longrightarrow [C_6H_6(H)(CH_3)]^+ \rightarrow \text{etc.}$$

$$C_6H_6 + {}^+C(=O)(CH_3){-}{-}{-}{-}{}^-AlCl_4 \longrightarrow [C_6H_6(H)(COCH_3)]^+ \rightarrow \text{etc.}$$

You should be familiar here with the full details of the mechanism, whether there is likely to be polysubstitution, etc. [See also sections 1.2(d), 1.4, 4.4(d), (e), 9.3(b).]

Friedel–Crafts reactions are useful for ring-building syntheses.

e.g.

$$C_6H_6 + (CH_2CO)_2O \xrightarrow{AlCl_3} C_6H_5-C(=O)-CH_2CH_2COOH \xrightarrow[HCl]{Zn/Hg} C_6H_5-CH_2CH_2CH_2COOH$$

$$\xrightarrow{SOCl_2} C_6H_5-CH_2CH_2CH_2COCl \xrightarrow{AlCl_3} \text{1-tetralone}$$

There are two Friedel–Crafts acylations involved here. The second is straightforward (though intramolecular); the first one also follows the mechanism we are familiar with since succinic anhydride (butanedioic anhydride) and $AlCl_3$ complex together to give essentially an acylium ion:

$$\text{succinic anhydride}\cdots AlCl_3 \longleftrightarrow CH_2(C^+{=}O)-CH_2-C(=O)-O-\bar{A}lCl_3$$

$$\text{(compare: } CH_3-\overset{O}{\overset{\|}{C}}-Cl\text{----}AlCl_3 \longleftrightarrow CH_3-\overset{O}{\overset{\|}{C}}{}^+ \quad \bar{A}lCl_4)$$

Note: Before carrying out the second Friedel–Crafts reaction we have to reduce the carbonyl group. This is because the ring closure must occur *ortho* to the first position of substitution and the carbonyl-containing substituent is *meta*-directing (and deactivating). [See section 10.7(e).]

(c) Conversion of diazonium salts to nitriles

One route leading to a new C—C bond involving a benzene ring is the following:

$$C_6H_6 \xrightarrow[\text{Conc. } H_2SO_4]{\text{Conc. } HNO_3} C_6H_5NO_2 \xrightarrow[HCl]{Sn} C_6H_5NH_2 \xrightarrow[HCl]{NaNO_2} C_6H_5\overset{+}{N}{\equiv}N\ Cl^- \xrightarrow{CuCN} C_6H_5CN$$

The mechanism of the last step (conversion of a benzenediazonium ion to a nitrile, the Sandmeyer reaction) may involve the following steps which culminate in the nucleophilic attack of cyanide ion on the phenyl carbocation:

$$C_6H_5-\overset{+}{N}{\equiv}N + Cu^+ \longrightarrow C_6H_5-N{=}N\cdot + Cu^{++}$$

$$C_6H_5-N{=}N\cdot \longrightarrow C_6H_5\cdot + N_2$$

$$C_6H_5\cdot + Cu^{++} \longrightarrow C_6H_5^+ + Cu^+$$

$$C_6H_5^+ + {}^-CN \longrightarrow C_6H_5-C{\equiv}N$$

(Diazonium salts are also highly useful for conversion into all sorts of other substituted benzenes.)

6.4 Nucleophilic attack on alkyl halides

The C—C bond-forming methods in this section are all basically S_N2 reactions [see section 5.4] and as such are all most successful when the alkyl halide is primary. In general we have:

$$-\overset{|}{\underset{|}{C}}{}^{-} \curvearrowright \overset{\delta+}{CH_2}(R)-\overset{\delta-}{X}$$

(a) Formation of nitriles

e.g.

$$NC^- \curvearrowright CH_2(CH_2CH_3)-Br \longrightarrow CH_3CH_2CH_2-C\equiv N + Br^-$$

Nitriles are useful synthetic intermediates en route to carboxylic acids and 1° amines:

$$R-C\equiv N \xrightarrow[\text{heat}]{H^+/H_2O} R-COOH \quad \text{[section 5.5(x)]}$$

$$R-C\equiv N \xrightarrow{LiAlH_4} RCH_2NH_2$$

(b) Reaction with alkynides

e.g.

$$CH_3CH_2C\equiv C^- \curvearrowright CH_2(CH_2CH(CH_3)_2)-Br \xrightarrow{\text{liq. }NH_3} CH_3CH_2C\equiv CCH_2CH_2CH(CH_3)_2 + Br^-$$

(c) Corey–House alkane synthesis [see also section 5.4(l)]

e.g.

$$(CH_3CH_2)_2CuLi + CH_3CH_2CH_2I \rightarrow CH_3CH_2CH_2CH_2CH_3$$

As with other organometallic reagents, the lithium dialkyl cuprate may be looked on as a source of R^- (here $CH_3CH_2^-$), although the carbanion is not actually free as such (compare Grignard reagents).

(d) Alkylation of ethyl acetoacetate (ethyl 3-oxobutanoate)

[See also sections 2.10, 5.4(k).]

e.g.

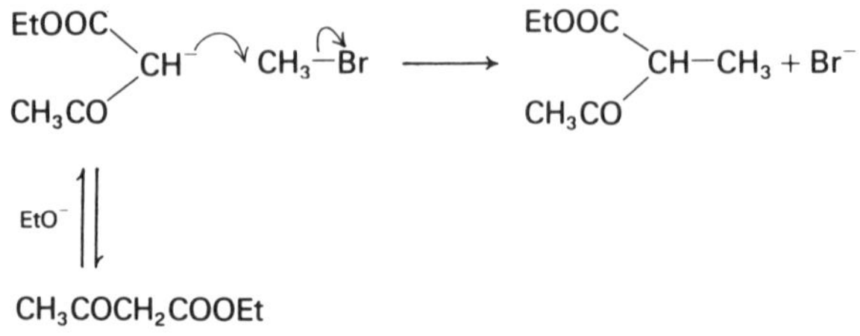

$$CH_3COCH_2COOEt \underset{}{\overset{EtO^-}{\rightleftharpoons}} (EtOOC)(CH_3CO)CH^- \;\; CH_3\text{—}Br \longrightarrow (EtOOC)(CH_3CO)CH\text{—}CH_3 + Br^-$$

The process may be repeated:

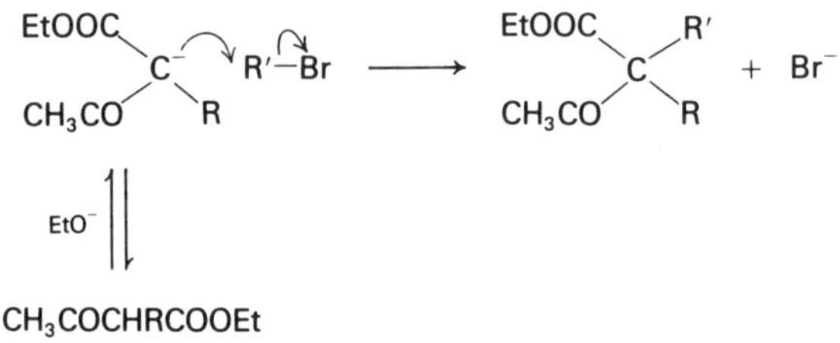

$$CH_3COCHRCOOEt \overset{EtO^-}{\rightleftharpoons} (EtOOC)(CH_3CO)C^-R \;\; R'\text{—}Br \longrightarrow (EtOOC)(CH_3CO)CRR' + Br^-$$

Upon hydrolysis and decarboxylation we end up with a ketone—the whole sequence of reactions is in fact a useful **ketone synthesis**.

e.g.

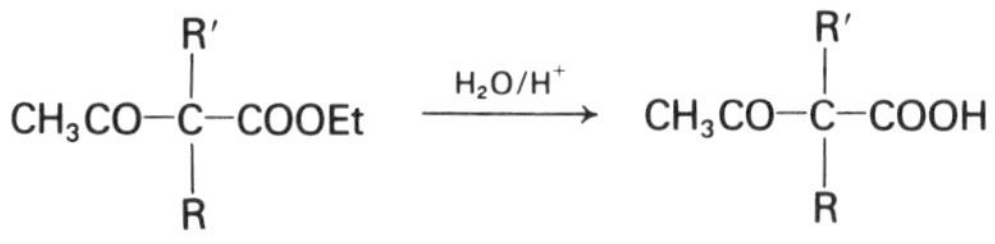

$$CH_3CO\text{—}CRR'\text{—}COOEt \xrightarrow{H_2O/H^+} CH_3CO\text{—}CRR'\text{—}COOH$$

Then:

$$CH_3\text{—}C(=O)\text{—}CRR'\text{—}C(=O)OH \xrightarrow[(-CO_2)]{heat} CH_3\text{—}C(OH)\text{=}CRR' + O\text{=}C\text{=}O$$

(an enol)

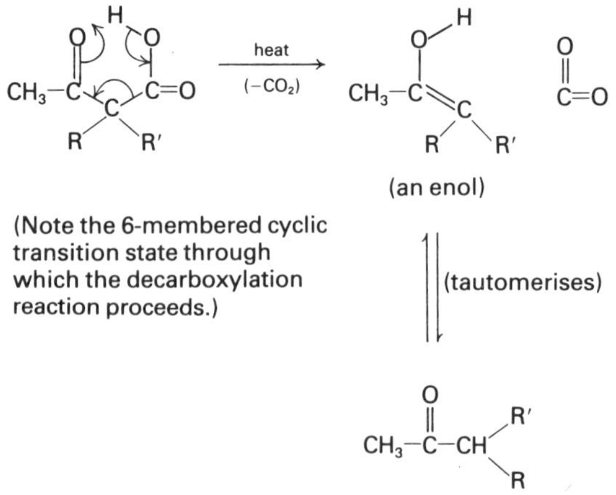

(Note the 6-membered cyclic transition state through which the decarboxylation reaction proceeds.)

(tautomerises)

$$CH_3\text{—}C(=O)\text{—}CHRR'$$

Different alkylations of ethyl acetoacetate involve using such things as Br—CH_2COOEt and Br—CH_2CO—R in place of a simple alkyl halide such as Br—CH_3. The mechanism involved in these cases is exactly the same, only the products are different. For example with an α-bromoester:

$$\begin{matrix} EtOOC \\ CH_3CO \end{matrix}\!\!>\!CH_2 \xrightarrow[\text{(ii) } BrCH_2COOEt]{\text{(i) NaOEt}} \begin{matrix} EtOOC \\ CH_3CO \end{matrix}\!\!>\!CH{-}CH_2COOEt$$

hydrolysis and decarboxylation of the β-ketoacid

$$\downarrow$$

$$CH_3COCH_2{-}CH_2COOH$$
(a γ-ketoacid)

(**Note:** It is the —COOH group in the β-position that decarboxylates. That in the γ-position is too far away to enter into the six membered cyclic mechanism that operates in the decarboxylation reaction.) And with an α-bromoketone:

$$\begin{matrix} EtOOC \\ CH_3CO \end{matrix}\!\!>\!CH_2 \xrightarrow[\text{(ii) } BrCH_2COR]{\text{(i) NaOEt}} \begin{matrix} EtOOC \\ CH_3CO \end{matrix}\!\!>\!CH{-}CH_2COR$$

hydrolysis and decarboxylation of the β-ketoacid

$$\downarrow$$

$$CH_3COCH_2{-}CH_2COR$$
(a γ-diketone)

(e) Alkylation of diethyl malonate (diethyl propanedioate)

e.g.

$$CH_2(COOEt)_2 \underset{}{\overset{EtO^-}{\rightleftharpoons}} \begin{matrix} EtOOC \\ EtOOC \end{matrix}\!\!>\!CH^- \;\; \underset{CH_3}{CH_2{-}I} \longrightarrow \begin{matrix} EtOOC \\ EtOOC \end{matrix}\!\!>\!CH{-}CH_2{-}CH_3 + I^-$$

Again, we can if necessary repeat the procedure to give:

$$\begin{matrix} EtOOC \\ EtOOC \end{matrix}\!\!>\!C\!<\!\begin{matrix} R' \\ R \end{matrix}$$

And then hydrolysis and decarboxylation as before lead to:

$$(EtOOC)_2C(R')(R) \xrightarrow{H_2O/H^+} (HOOC)_2C(R')(R) \xrightarrow[-CO_2]{heat} HOOC{-}CH(R')(R)$$

This time the sequence of reactions constitutes a useful **carboxylic acid synthesis**. (Instead of diethyl malonate one could start here with ethyl cyanoacetate, N≡C—CH_2—COOEt, and end up with the same carboxylic acid product. Can you follow the mechanism of each step that would be involved here?)

6.5 Nucleophilic attack on carbonyl groups

This section constitutes an extremely important and versatile way of making new C—C bonds. As you go through the following examples remember that with aldehydes and ketones we are seeing nucleophilic addition [section 5.6] whereas with acyl compounds we have nucleophilic attack followed by the expulsion of a leaving group [section 5.5]. In other words, for each of the reactions in this section, decide which of the following general reaction schemes is involved:

$$-\overset{|}{\underset{|}{C}}{}^{-} \;+\; \overset{|}{\underset{|}{C}}{=}O \longrightarrow -\overset{|}{\underset{|}{C}}-\overset{|}{\underset{|}{C}}-O^- \xrightarrow{H^+} -\overset{|}{\underset{|}{C}}-\overset{|}{\underset{|}{C}}-OH$$

$$-\overset{|}{\underset{|}{C}}{}^{-} \;+\; \overset{X}{\underset{|}{C}}{=}O \longrightarrow -\overset{|}{\underset{|}{C}}-\overset{X}{\underset{|}{C}}-O^- \longrightarrow -\overset{|}{\underset{|}{C}}-C{=}O + X^-$$

(a) Reaction of Grignard reagents with aldehydes and ketones

In general:

$$-\overset{|}{\underset{|}{C}}{}^{-} \;+\; \overset{A}{\underset{B}{C}}{=}O \longrightarrow -\overset{|}{\underset{|}{C}}-\overset{A}{\underset{B}{C}}-O^- \;\; \overset{+}{Mg}X \xrightarrow[H^+]{H_2O} -\overset{|}{\underset{|}{C}}-\overset{A}{\underset{B}{C}}-OH$$

(i.e. $R^{\delta-}$—MgX)

(We think of the Grignard reagent as supplying a carbanion.) The product is an alcohol. If A = B = H (i.e. we start with formaldehyde, or methanal) the product is a 1° alcohol; if A = H, B = R the product is 2°, and if A = R, B = R′ the product is 3°. Thus we can synthesise an enormous range of alcohols using Grignard reagents, often in more than one way.

e.g.

$$C_6H_5MgBr + H_2C{=}O \longrightarrow C_6H_5CH_2OH$$

$$CH_3CH_2MgCl + CH_3CH_2CHO \rightarrow (CH_3CH_2)_2CHOH$$

$$C_6H_5MgBr + CH_3COCH_2CH_3$$

$$CH_3MgBr + C_6H_5COCH_2CH_3 \longrightarrow C_6H_5C(CH_3)(OH)CH_2CH_3$$

$$CH_3CH_2MgBr + C_6H_5COCH_3$$

(b) Reaction of Grignard reagents with esters

[See also section 5.5(r).] This is an alternative way of synthesising 3° alcohols (with at least two of the three R groups the same).

e.g.

$$CH_3COOEt \xrightarrow{PhMgBr} \left[CH_3{-}CO{-}Ph\right] \xrightarrow{PhMgBr} CH_3{-}C(OH)(Ph){-}Ph$$

(not isolated)

(c) Reaction of Grignard reagents with carbon dioxide

e.g.

$$CH_3{-}C(CH_3)_2{-}MgBr \xrightarrow[\text{(ii) } H^+]{\text{(i) } CO_2} CH_3{-}C(CH_3)_2{-}COOH$$

The alternative route to this particular acid, involving R—Br→R—CN→RCOOH, is not possible. [Why not? See section 7.7(b).] In general, these two methods of carboxylic acid synthesis enable an alcohol to be turned into the *next higher* carboxylic acid, whereas simple oxidation of course leads to a product with the *same* number of carbon atoms:

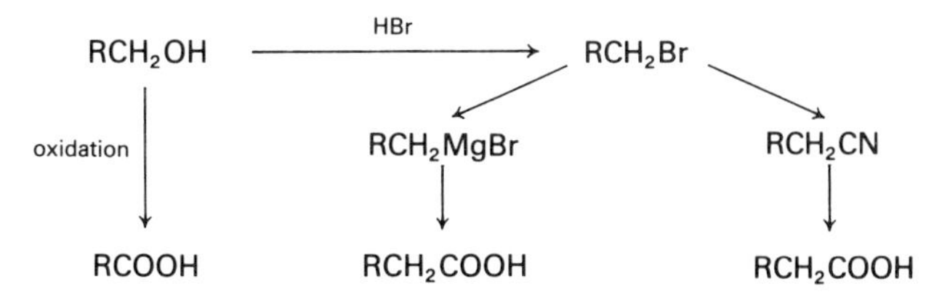

(d) Reaction of Grignard reagents with ethylene oxide (oxirane)

[See section 5.4(o).] This provides a way of increasing the chain length of an alcohol by two carbons:

$$R{-}OH \xrightarrow{HBr} R{-}Br \xrightarrow{Mg} RMgBr \xrightarrow{CH_2{-}CH_2 \text{ (O bridged)}} RCH_2CH_2OH$$

(e) Nucleophilic attack by phenoxide ion

(Before continuing with other organometallic reagents we will make a brief detour to look at a couple of reactions of the phenoxide ion in which the products are similar to those obtained in a Grignard reaction.)

We would usually expect a phenoxide ion to atack a carbonyl group as follows (e.g. in the preparation of phenyl esters [section 5.5(c)]):

$$C_6H_5{-}O^- + {>}C{=}O \longrightarrow C_6H_5{-}O{-}C{-}O^- \longrightarrow \text{etc.}$$

(This does not lead to a new C—C bond.) On occasions, however, an alternative mode of attack is possible:

$$C_6H_5O^- + {>}C{=}O \longrightarrow \text{(cyclohexadienone with H and } {-}C{-}O^- \text{ on the 2-position)}$$

This might be easier to see if you bear in mind that other resonance forms of the phenoxide ion have the negative charge on the benzene ring [see section 10.2(b)]. That is, we could write:

$$\text{(cyclohexadienone carbanion)} + {>}C{=}O \longrightarrow \text{(cyclohexadienone with H and } {-}C{-}O^- \text{ on the 2-position)}$$

Two examples of this process are the reactions with *carbon dioxide* and with *formaldehyde*. The first of these is the Kolbé–Schmitt method for the synthesis of salicylic acid (2-hydroxybenzoic acid):

$$C_6H_5O^- + O{=}C{=}O \longrightarrow \text{(cyclohexadienone with H and } {-}COO^-) \xrightarrow{-H^+} \text{(}O^- \text{, } COO^- \text{ on ring)} \xrightarrow{2H^+} \text{(OH, COOH on benzene ring)}$$

The second involves the following reaction (and is the first step in the preparation of **Bakelite**):

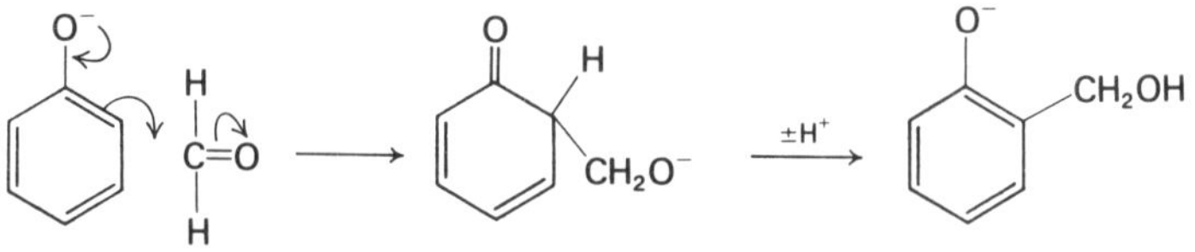

O^- H C=O H → O H CH_2O^- $\xrightarrow{\pm H^+}$ O^- CH_2OH

Connection In both these reactions the phenoxide ion is acting like a **carbanion** in the same way as a Grignard reagent does (and in fact is giving the same kinds of products as we have just seen would arise with a Grignard reagent—namely a carboxylic acid and a 1° alcohol, respectively.

Note: (i) As always with this type of reaction we can describe it in two ways—as nucleophilic attack of the phenoxide ion on the δ+ charge of the carbonyl group, or as electrophilic attack of the carbonyl compound on the electrons of the aromatic ring.

(ii) In the second reaction we can also get substitution at the other ortho and para positions; this is then followed by cross-linking reactions such as:

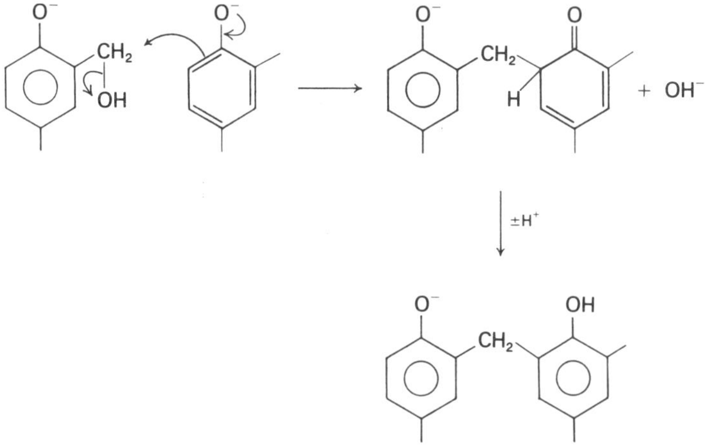

O^- CH_2 OH O^- → O^- CH_2 O H + OH^-

$\downarrow \pm H^+$

O^- CH_2 OH

The final product is a highly cross-linked, hard, brittle polymer (Bakelite) with this kind of structure (which you should try to imagine as extending in three dimensions, not just two, but that would be difficult to draw!):

(f) Synthesis of ketones via organometallic reagents

Organocadmiums and lithium organocuprates can usefully convert acyl chlorides to ketones.

e.g.

$$CH_3CH_2-\overset{O}{\overset{\|}{C}}-Cl \xrightarrow{[(CH_3)_2CH]_2Cd} CH_3CH_2-\overset{O}{\overset{\|}{C}}-CH(CH_3)_2$$

$$O_2N-C_6H_4-COCl \xrightarrow{(CH_3)_2CuLi} O_2N-C_6H_4-COCH_3$$

Remember, a Grignard reagent would react further, converting the ketone to a 3° alcohol.

(g) Reformatsky reaction

[See also sections 2.11, 10.3(c).]

e.g.

$$\text{1-methyl-1-(COOCH}_3\text{)-cyclopentan-3-one} \xrightarrow[\text{(ii) } H^+/H_2O]{\text{(i) Zn, BrCH}_2\text{COOEt}} \text{1-methyl-1-(COOCH}_3\text{)-3-(OH)-3-(CH}_2\text{COOEt)-cyclopentane}$$

Note: (i) The reactive species is an organozinc reagent.
(ii) It is *like* a Grignard reagent in that it converts a ketone group to a 3° alcohol.
(iii) It is *unlike* a Grignard reagent in that it does not react with ester groups (either in itself or in the other reactant).

(h) The aldol condensation

In section 5.6(e) we exemplified this reaction as follows:

$$2CH_3CH_2CHO \xrightarrow{\text{dil. } OH^-} CH_3CH_2{-}CH(OH){-}CH(CH_3){-}CHO$$

If this aldol is in fact the product, we should call the reaction an aldol **addition**. If, however, the product spontaneously loses water as it easily can do [to form in this case $CH_3CH_2CH{=}C(CH_3)CHO$] then we are properly justified in calling it an aldol **condensation**. ('Condensation' implies the loss of a small molecule such as water.) Loosely though, the term 'aldol condensation' is often used for this type of reaction regardless of whether the product loses water or not.

With ketones the analogous reaction is often less successful, although with the right conditions a good yield of product can be obtained in simple cases.

e.g.

$$CH_3{-}C({=}O){-}CH_3 \underset{}{\overset{\text{base}}{\rightleftharpoons}} CH_3{-}C({=}O){-}CH_2^- \longleftrightarrow CH_3{-}C(O^-){=}CH_2$$

$$CH_3{-}C({=}O){-}CH_2^- + (CH_3)_2C{=}O \rightleftharpoons CH_3{-}C({=}O){-}CH_2{-}C(CH_3)_2{-}O^- \xrightarrow{H^+} CH_3{-}C({=}O){-}CH_2{-}C(CH_3)_2{-}OH$$

(Strictly speaking, the product is not an aldol, but a 'ketol'.)

(i) Crossed aldol condensation

If an aldehyde has no α-hydrogen it cannot of course condense with itself. But it can condense with an aldehyde that does possess an α-hydrogen in a 'crossed aldol' reaction.

e.g.

$$C_6H_5{-}CHO + CH_3CHO \xrightarrow{OH^-} C_6H_5{-}CH{=}CH{-}CHO$$

In this synthesis benzaldehyde is first mixed with dilute sodium hydroxide solution. No aldol reaction can happen. Ethanal (acetaldehyde) is then dripped in slowly. As soon as each enolate ion is formed by abstraction of a proton from the ethanal molecule, it finds itself in the presence of a large excess of benzaldehyde—so it condenses with that rather than condensing with a second molecule of ethanal, which is present in much lower concentration. (Can you write out the mechanism for each step of the reaction?)

(j) Aldol-type cyclisations

Building up ring systems is important in the synthesis of many organic compounds, such as steroids for example. Robinson utilised the aldol-type mechanism for syntheses such as:

(Can you recognise that the mechanism here is exactly the same as in the simple dimerisation of acetone shown in section 6.5(h)?)

(k) The Claisen condensation

[See also sections 2.11, 5.5(p), 7.6(c).]

e.g.

$$CH_3CH_2COOEt \xrightarrow{EtO^-} CH_3CH_2{-}\overset{O}{\overset{\|}{C}}{-}\underset{CH_3}{\underset{|}{CH}}{-}\overset{O}{\overset{\|}{C}}{-}OEt$$

Recall the mechanism and how similar it is to the aldol condensation, except for the loss of a leaving group:

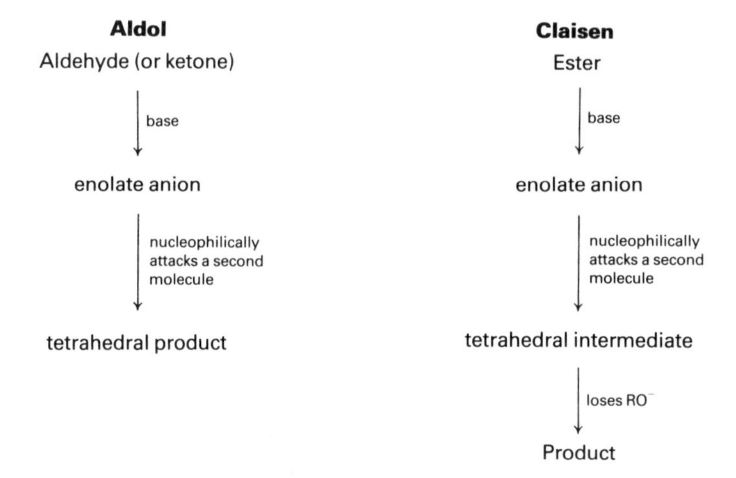

(l) Crossed Claisen condensations

Just as crossed aldol condensations are possible when one reactant lacks α-hydrogens, so crossed Claisen condensations can arise under similar circumstances.

e.g.

$$CH_3-C_6H_4-COOEt + CH_3COOEt \xrightarrow[\text{(ii) } H^+]{\text{(i) } EtO^-} CH_3-C_6H_4-\overset{O}{\overset{\|}{C}}-CH_2-\overset{O}{\overset{\|}{C}}-OEt$$

(no α-hydrogen)

(Try writing out the full mechanism.)

(m) Claisen-type cyclisations

Like the aldol reaction, the Claisen has also been adapted for the synthesis of alicyclic compounds.

e.g.

$EtOOC-CH_2-CH_2-CH_2-CH_2-COOEt \underset{}{\overset{EtO^-}{\rightleftharpoons}}$ $EtOOC-\overset{-}{C}H-CH_2-CH_2-CH_2-C(=O)OEt$ $\rightleftharpoons$ (cyclic tetrahedral intermediate: CH_2, CH_2, $CH_2-CH(COOEt)$, $C(O^-)(OEt)$) $\rightarrow$ 2-oxocyclopentane with H and COOEt

This is called the Dieckmann reaction, but I hope you can see that the mechanism is exactly the same as for the Claisen.

The hydrogen drawn in the product is relatively acidic for the same reason as in ethyl 3-oxobutanoate (i.e. it is alpha to two carbonyl groups). Therefore the product can be alkylated and then hydrolysed and decarboxylated in just the same way as we have seen in sections 5.4(k), 6.4(d):

O, H, COOEt → (i) EtO^- (ii) R–Br → O, R, COOEt → H_2O, H^+ → O, R, COOH (a β-ketoacid) → heat → O, R, H

This sequence constitutes a synthesis of variously substituted cyclopentanones. Analogous reactions can lead to substituted cyclohexanones.

(n) Acylation of β-ketoesters

We have seen that β-ketoesters can be **alkylated** [section 5.4(k)]. In a similar way they can be **acylated**. (The reaction must be done in a solvent that itself does not react with the acyl halide, of course.)

e.g.

$$CH_3COCH_2COOEt \xrightarrow{\text{base}} (CH_3CO)(EtOOC)CH^- + Cl\text{–}C(=O)\text{–}CH_3 \longrightarrow (CH_3CO)(EtOOC)CH\text{–}C(Cl)(CH_3)\text{–}O^- \longrightarrow (CH_3CO)(EtOOC)CH\text{–}C(=O)\text{–}CH_3 + Cl^-$$

Then in this case the usual sequence of hydrolysis and decarboxylation leads to a β-diketone:

$$(CH_3CO)(EtOOC)CH\text{–}COCH_3 \longrightarrow (CH_3CO)(HOOC)CH\text{–}COCH_3 \xrightarrow{-CO_2} CH_3COCH_2COCH_3$$

pentane-2,4-dione (acetylacetone)

(o) Knoevenagel and Doebner reactions

We have seen that the anions derived from such compounds as ethyl acetoacetate (3-oxobutanoate) and diethyl malonate (propanedioate) can displace halide from alkyl halides [section 5.4(k)] and from acyl halides [preceding section]. From what we know of carbanions it should be no surprise to learn that such anions can also add nucleophilically to aldehydes and ketones.

e.g.

$$(EtOOC)(CH_3CO)CH^- + C_6H_5CH{=}O \rightleftharpoons (EtOOC)(CH_3CO)CH{-}CH(C_6H_5){-}O^- \xrightarrow{H^+} (EtOOC)(CH_3CO)CH{-}CH(C_6H_5){-}OH$$

$$CH_3COCH_2COOEt \xrightarrow{base} (EtOOC)(CH_3CO)CH^-$$

$$(EtOOC)(CH_3CO)CH{-}CH(C_6H_5){-}OH \xrightarrow{-H_2O} C_6H_5{-}CH{=}C(COOEt)(COCH_3)$$

(The first formed product loses water readily because the resulting double bond is **conjugated** with the benzene ring and the carbonyl groups.) Reactions such as these are known as Knoevenagel reactions. If one of the starting materials is malonic (propanedioic) acid, the product often spontaneously decarboxylates and then the reaction is called a Doebner reaction.

e.g.

$$(HOOC)_2CH^- + C_6H_{13}CH{=}O \longrightarrow (HOOC)_2CH{-}CH(C_6H_{13}){-}O^- \xrightarrow{H^+} (HOOC)_2CH{-}CH(C_6H_{13}){-}OH$$

$$CH_2(COOH)_2 \xrightarrow{base\ (e.g.\ pyridine)} (HOOC)_2CH^-$$

$$(HOOC)_2CH{-}CH(C_6H_{13}){-}OH \xrightarrow{-H_2O} C_6H_{13}{-}CH{=}C(COOH)_2 \xrightarrow{-CO_2} C_6H_{13}{-}CH{=}CH{-}COOH$$

If we have an α,β-unsaturated aldehyde or ketone then sometimes the nucleophile attacks predominantly at the β-position as we have seen [in section 5.8].

It is not important that you memorise the names associated with some of these reactions (e.g. Knoevenagel, Doebner, Dieckmann—and also Perkin and

Stobbe, whose reactions we have not covered but which you might care to look into). The important thing is to realise how similar they all are to each other and to the aldol condensation. They all involve generation of a resonance-stabilised anion (i.e. an enolate ion) which then nucleophilically attacks the carbonyl group of an aldehyde, ketone, ester, etc. A possible major divergence in the mechanism is whether or not the resultant tetrahedral species can expel a leaving group.

(p) Wittig reactions

[See also sections 2.11, 5.6(f).] The first step in a Wittig reaction is nucleophilic attack on a carbonyl group (as it is for all the C—C bond-forming reactions in this section):

$$-\overset{|}{C}^{-}(-P^{+}-) \; + \; \overset{|}{C}{=}O \longrightarrow -\overset{|}{C}-\overset{|}{C}-O^{-}(-P^{+}-) \xrightarrow[\text{steps}]{\text{several}} \; >C{=}C< \; + \; O{=}P-$$

$$-\overset{|}{C}^{-}-P^{+}- \;\longleftrightarrow\; >C{=}P-$$

The end result represents an important mode of synthesis of carbon–carbon **double** bonds, e.g.

$$C_6H_5-CH{=}CH-\overset{-}{C}H-\overset{+}{P}(Ph)_3 \longleftrightarrow C_6H_5-CH{=}CH-CH{=}P(Ph)_3$$

$$+ \; C_6H_5-CHO \longrightarrow C_6H_5-CH{=}CH-CH{=}CH-C_6H_5$$

(q) Summary

I cannot stress too much how important the synthetic reactions in this section are. Grignard, aldol, Claisen, Wittig—these types of reaction are some of the most valuable in the armoury of the practising synthetic organic chemist. They all involve attack of a carbon nucleophile on a compound containing a carbonyl group.

6.6 Nucleophilic attack on carbenes

Carbenes are reactive, short-lived, **divalent** compounds of carbon, such as CH_2 or CCl_2. Since the carbon has only six valence electrons, such species are electrophilic. If they react with carbon nucleophiles then new C—C bonds are formed.

e.g.

$$>C{=}C< + CH_2 \longrightarrow \text{cyclopropane (C–C with bridging } CH_2)$$

Compare the electrophilic attack of bromine:

$$>C{=}C< + Br{-}Br \longrightarrow \text{bromonium ion } (C{-}C \text{ bridged by } Br^+) + Br^-$$

Dichlorocarbene is involved in the **Reimer–Tiemann** reaction, the reaction of phenoxide ion with chloroform (trichloromethane) and sodium hydroxide solution to give 2-hydroxybenzaldehyde (salicylaldehyde):

$$CHCl_3 + OH^- \rightleftharpoons Cl_3C^- + H_2O$$

$$Cl_3C^- \rightarrow CCl_2 + Cl^-$$

Then:

phenoxide + CCl_2 ⟶ (dienone with H and CCl_2^-) $\xrightarrow{\pm H^+}$ phenoxide with $CHCl_2$

↓ alkaline hydrolysis of a *gem*-dihalide

phenoxide with CHO $\xrightarrow{H^+}$ OH/CHO

salicylaldehyde
(2-hydroxybenzenecarbaldehyde)

Does the attack of phenoxide ion on CCl_2 remind you of anything? It should—it is very similar to the attack of phenoxide ion on CO_2 and on HCHO [section 6.5(e)]. In all these cases benzene itself would not react; only the highly activated (that is, more nucleophilic) phenoxide ion is capable of reaction. Exactly the same difference of reactivity has also been noted between benzene and the phenoxide ion with respect to the electrophilic attack of a diazonium ion [section 4.5].

6.7 Diels–Alder reactions

There are several types of cyclisation reaction for which it is difficult to represent the mechanism in simple 'arrow pushing' terms, but which can be thoroughly understood in terms of modern Molecular Orbital theory as expounded by Woodward and Hoffmann (beyond our scope here). One of these is the Diels–Alder reaction; basically this is the combination of a diene and an alkene to give a cyclohexene:

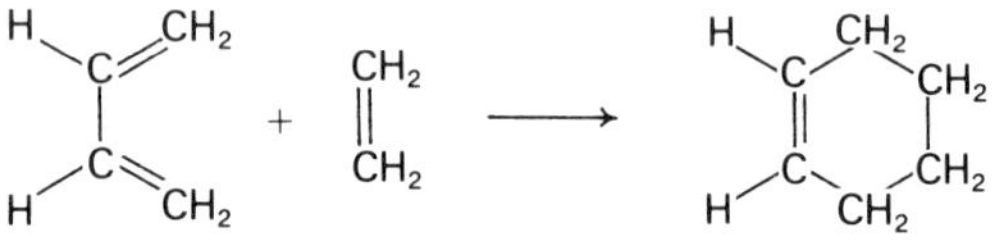

The alkene (known as the dienophile because it 'loves' to react with the diene) is generally not just a simple alkene such as ethene, but is one that carries one or more electron-withdrawing groups. Sometimes the dienophile is a substituted alkyne rather than an alkene. Examples of dienophiles are:

$CH_2{=}CH{-}CHO$, (maleic anhydride) , $(NC)_2C{=}C(CN)_2$, $EtOOC{-}C{\equiv}C{-}COOEt$

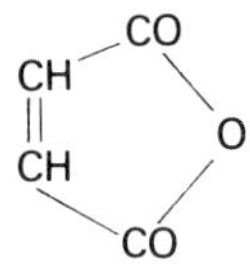

The diene also may be of considerably more complex structure than the simplest case, 1,3-butadiene. Examples of Diels–Alder reactions are the following, showing some of the various substituted cyclohexenes that may be produced by this versatile C—C bond-forming reaction:

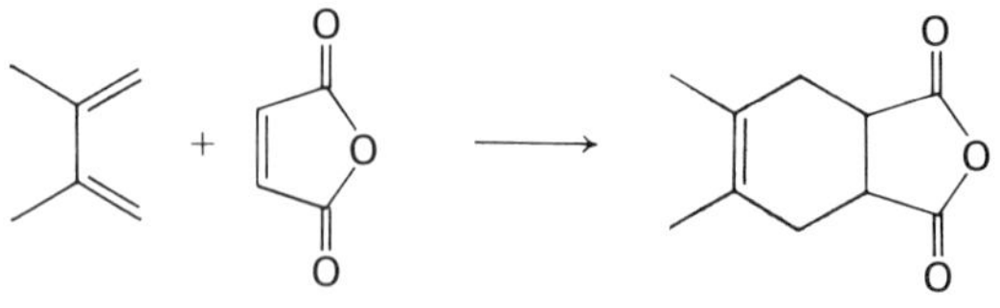

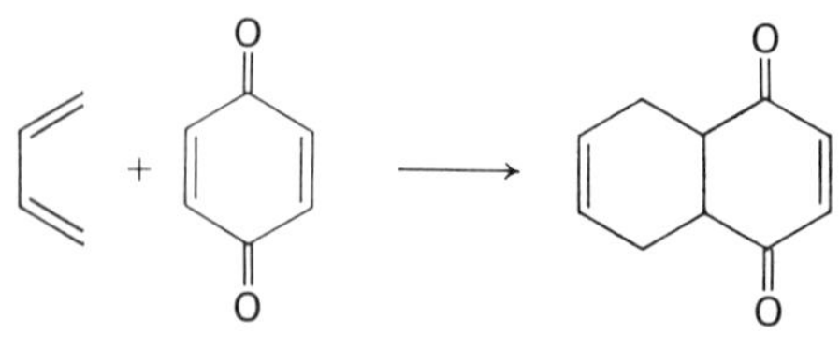

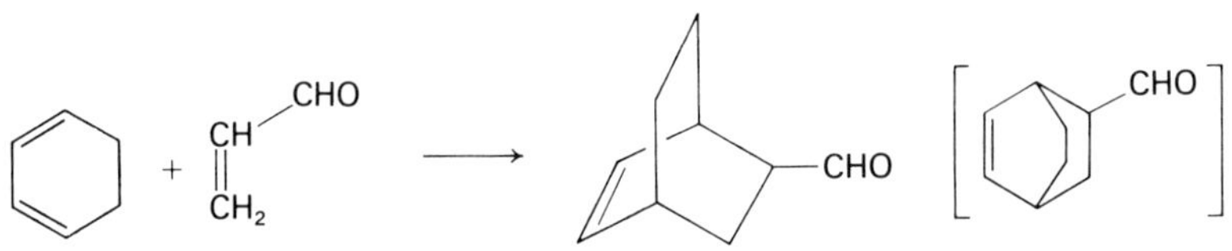

CHO
CH
CH$_2$
CHO
CHO

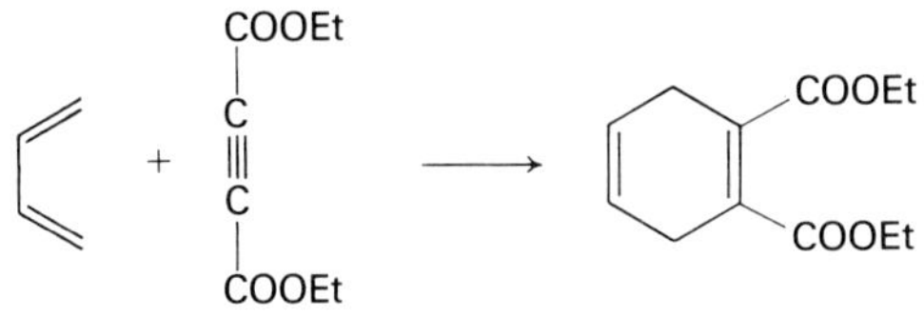

COOEt
C
C
COOEt
COOEt
COOEt

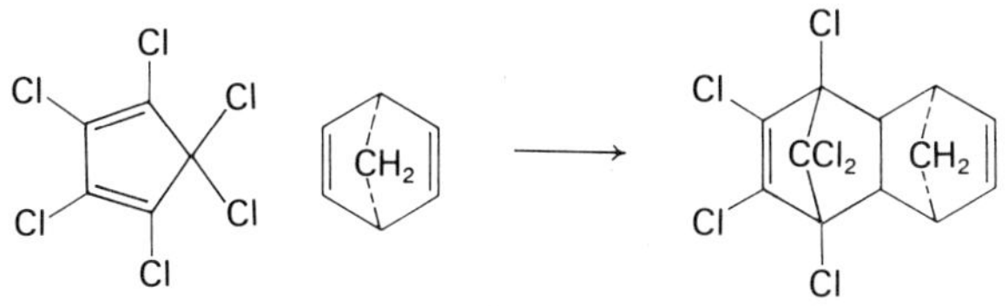

Cl Cl Cl Cl Cl Cl
CH$_2$
Cl Cl Cl Cl Cl
CCl$_2$ CH$_2$

The last product above is 'Aldrin'—at one time a popular insecticide. 'Dieldrin' is the same but with the right hand double bond turned into an epoxide.

7

Acid/Base Reactions

7.1 Introduction

We have seen that whenever something acts as an electrophile, whatever it reacts with must be a nucleophile. In a similar way, we will see in this chapter that whenever something acts as an acid, whatever it reacts with must be a base. (A third pair that always go together are oxidation and reduction; whenever something acts as an oxidising agent, whatever it reacts with is a reducing agent.)

In the vast majority of acid/base reactions in organic chemistry a **proton** is transferred from the acid to the base. On the Lowry–Bronsted system, an **acid** is defined as a substance capable of **donating** a proton, and a **base** is a substance that can **accept** a proton.

A broader definition has been suggested by Lewis. A **Lewis acid** is a substance capable of **accepting a pair of electrons**, whereas a **Lewis base** is one that can **donate a pair of electrons**. You will recall that these definitions are exactly those we have already met for electrophiles and nucleophiles respectively:

An acceptor of an electron-pair	=	an electrophile	=	a Lewis acid
A donator of an electron-pair	=	a nucleophile	=	a Lewis base

A proton, which is so important in the Lowry–Bronsted description of acids and bases, is just one particular example of a Lewis acid. Clearly, when a proton reacts with a base it is **accepting an electron-pair**:

$$B\!:\, + H^{+} \rightarrow \overset{+}{B}\!\div\!H$$

On the Lewis definition, all the multitudinous reactions between electrophiles and nucleophiles that we have already looked at could also be classified as acid–base reactions. However, it is more convenient to restrict ourselves to acid–base reactions following the Lowry–Bronsted description (i.e. those involving proton transfer) and it is this meaning of 'acid' and 'base' that is used in this chapter.

7.2 Relative strengths of acids and bases

An acid tends to donate a proton, by definition. The stronger this tendency, the stronger the acid. What is left when the proton has been given away is called the 'conjugate base'. If the proton is lost readily, it must mean that the conjugate base is only a weak base, otherwise it would strongly resist the loss of the proton. In other words: **the conjugate base of a strong acid is a weak base**, and by the same token: **the conjugate base of a weak acid is a strong base**.

A quantitative measure of the strength of an acid is derived from its tendency to donate a proton to water:

$$HA + H_2O \rightleftharpoons H_3O^+ + A^-$$

$$K = \frac{[H_3O^+][A^-]}{[HA][H_2O]}$$

$$\therefore K[H_2O] = \frac{[H_3O^+][A^-]}{[HA]} = K_a$$

$$\text{and } pK_a = -\log_{10} K_a$$

$$\text{or } pK_a = \log_{10}\left(\frac{1}{K_a}\right)$$

Since pK_a is the log of the reciprocal of K_a (the acidity constant) it means that as K_a gets bigger, pK_a gets smaller.

> The stronger the acid, the larger the K_a.
> The stronger the acid, the smaller the pK_a.

From the expression above for pK_a, it is obvious that:

$$pK_a = -\log_{10}\left(\frac{[H_3O^+][A^-]}{[HA]}\right)$$

$$= -\log_{10}[H_3O^+] - \log_{10}\frac{[A^-]}{[HA]}$$

$$= pH - \log_{10}\frac{[A^-]}{[HA]}$$

If $[A^-] = [HA]$, then $\log[A^-]/[HA] = 0$, and $pK_a = pH$. In other words: **the pK_a is the pH at which an acid is half ionised**. At any pH greater than the pK_a there is more A^- than HA present, and at any pH lower than the pK_a there is more HA than A^- present.

Table 7.1 lists most of the common classes of organic compound together with their pK_a values and their conjugate bases, and also several inorganic acids for purposes of comparison. The acids at the top of the table are the weakest ones; correspondingly, the bases at the top are the strongest ones. *It is very useful indeed to be able to grasp the approximate whereabouts in this table of a particular organic acid.*

Table 7.1 The strengths of acids and bases

Acid	pK_a	Base
CH_3CH_3	~42	$CH_3CH_2^-$
H_2	~40	H^-
$CH_2{=}CH_2$	36	$CH_2{=}CH^-$
NH_3	34	NH_2^-
Ph_3CH	32	Ph_3C^-
$HC{\equiv}CH$	25	$HC{\equiv}C^-$
CH_3COOEt	25	$^-CH_2COOEt$
$CHCl_3$	25	$^-CCl_3$
CH_3CN	25	$^-CH_2CN$
CH_3COCH_3	20	$CH_3COCH_2^-$
EtOH	18	EtO^-
H_2O	15.7	HO^-
(cyclopentadiene, CH_2 with H, H)	15	[cyclopentadienyl]$^-$
$[(H_2N)_3C]^+$	13.6	$NH{=}C(NH_2)_2$ (guanidine)
EtSH	11–12	EtS^-
$MeCOCH_2COOEt$	10.7	$MeCO\bar{C}HCOOEt$
$PhSO_2NH_2$	10	$PhSO_2NH^-$
CH_3NO_2	10	$^-CH_2NO_2$
Ph–OH	9.9	PhO^-
$CH_3{-}NH_3^+$	9.64	$CH_3{-}NH_2$
NH_4^+	9.25	NH_3
HCN	9.1	CN^-
$MeCOCH_2COMe$	9.0	$MeCO\bar{C}HCOMe$
(phthalimide: C_6H_4(CO)$_2$NH)	9	C_6H_4(CO)$_2N^-$
H_2S	7	HS^-
H_2CO_3	6.5	HCO_3^-
MeCOOH	4.8	$MeCOO^-$
$Ph{-}NH_3^+$	4.6	$Ph{-}NH_2$
Ph–COOH	4.2	$PhCOO^-$
HF	3.2	F^-
$Ph{-}SO_3H$	2.6	$Ph{-}SO_3^-$
H_3PO_4	2.1	$H_2PO_4^-$
H_3O^+	−1.7	H_2O
$Me{-}C({=}\overset{+}{O}H){-}NH_2$	−1 to −2	$MeCONH_2$
$MeOH_2^+$	−2.2	MeOH
H_2SO_4	−3	HSO_4^-
$Et_2\overset{+}{O}H$	−3.6	Et_2O
$Me{-}C({=}\overset{+}{O}H){-}OH$	−6.1	MeCOOH

Increasing acidity (down the Acid column); Increasing basicity (up the Base column); Increasing leaving group ability (down the Base column)

Table 7.1 (cont.)

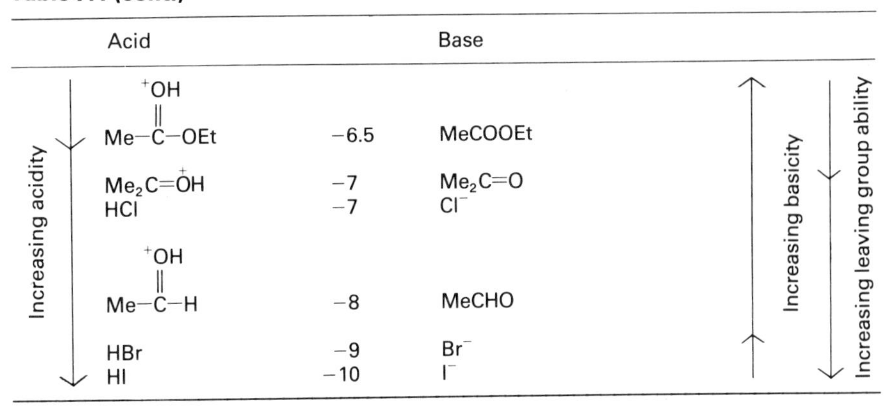

Acid	pK_a	Base
$Me-C(=\overset{+}{O}H)-OEt$	−6.5	MeCOOEt
$Me_2C=\overset{+}{O}H$	−7	$Me_2C=O$
HCl	−7	Cl^-
$Me-C(=\overset{+}{O}H)-H$	−8	MeCHO
HBr	−9	Br^-
HI	−10	I^-

Increasing acidity (down the acid column); Increasing basicity (up the base column); Increasing leaving group ability (down the base column)

Many compounds can act both as acids and as bases, and thus several of them appear twice in Table 7.1.

e.g.

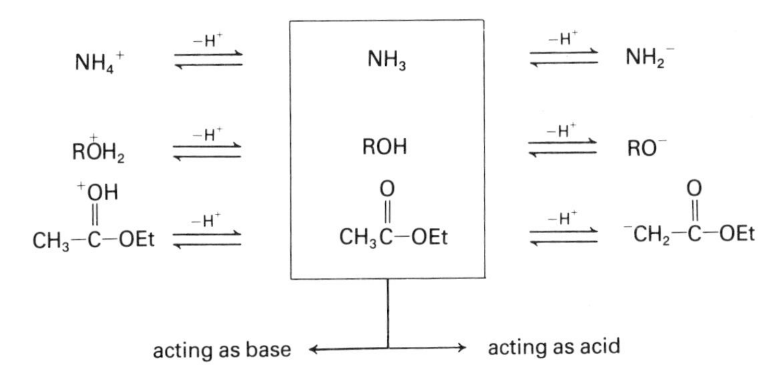

It is very important that you can identify the pK_a that refers to each reaction. For instance, the pK_a of NH_3 is 34 and this has *nothing to do with ammonia acting as a base*—it is a measure of the **acidity** of NH_3, its tendency to donate a proton and become NH_2^-. If it is the **basicity** of NH_3 you are interested in then this is defined by the pK_a of the conjugate acid, NH_4^+. For example, if you wanted to compare the basicity of NH_3 with that of CH_3NH_2 you would note that the pK_a of NH_4^+ is 9.25 and of $CH_3NH_3^+$ is 9.64. Thus NH_4^+ is a slightly stronger acid than $CH_3NH_3^+$ and therefore NH_3 is a slightly weaker base than CH_3NH_2.

7.3 Position of equilibrium in acid/base reactions

In general the exchange of a proton between an acid and a base is an equilibrium reaction:

$$HA + B^- \rightleftharpoons HB + A^-$$

In the forward direction HA is acting as an acid and B^- as a base; in the reverse direction HB is an acid and A^- is a base. The position of equilibrium depends upon the relative strengths of these species as acids and bases and this information can be obtained from Table 7.1 (In practice, of course, the process happens the other way—the pK_a data in the table are estimated from whether or not certain reactions occur.)

If HA is a stronger acid than HB then by definition B^- must be a stronger base than A^- and under these circumstances the position of equilibrium favours the products:

H–A	+ B$^-$	$\rightleftharpoons$	H–B	+ A$^-$
(stronger acid)	(stronger base)		(weaker acid)	(weaker base)

On the other hand, if HA is a weaker acid than HB then the position of equilibrium favours the reactants (that is, the reaction does not 'go', or at least not as much):

H–A	+ B$^-$	$\rightleftharpoons$	H–B	+ A$^-$
(weaker acid)	(weaker base)		(stronger acid)	(stronger base)

We can think of many cases of the first type simply by looking at the relative pK_as in Table 7.1. Several examples of this sort are listed in Table 7.2.
If there is a very large difference in pK_a between the pair of acids involved in one of these reactions then the position of equilibrium lies so overwhelmingly over to

Table 7.2

Stronger acid	Stronger base		Weaker acid	Weaker base
H_2O	+ $CH_3CH_2^-$	$\rightleftharpoons$	CH_3CH_3	+ OH^-
cyclopentadiene (ring with CH_2, H, H)	+ Ph_3C^-	$\rightleftharpoons$	Ph_3CH	+ cyclopentadienyl anion
R–C≡C–H	+ NH_2^-	$\rightleftharpoons$	NH_3	+ R–C≡C$^-$
H_3O^+	+ CH_3NH_2	$\rightleftharpoons$	$CH_3NH_3^+$	+ H_2O
H_2O	+ $CH_3CH_2O^-$	$\rightleftharpoons$	CH_3CH_2OH	+ OH^-
CH_3CH_2SH	+ OH^-	$\rightleftharpoons$	H_2O	+ $CH_3CH_2S^-$
PhOH	+ OH^-	$\rightleftharpoons$	H_2O	+ Ph–O$^-$
RCOOH	+ HCO_3^-	$\rightleftharpoons$	H_2CO_3 ($\rightarrow CO_2 + H_2O$)	+ $RCOO^-$
H_2O	+ Cl_3C^-	$\rightleftharpoons$	$CHCl_3$	+ OH^-
H_2O	+ $CH_3COCH_2^-$	$\rightleftharpoons$	CH_3COCH_3	+ OH^-
EtOH	+ $^-CH_2COOEt$	$\rightleftharpoons$	CH_3COOEt	+ EtO^-

the right that we can say to all intents and purposes the reaction goes to completion.

e.g.

$$H_2O + CH_3CH_2^- \rightarrow CH_3CH_3 + OH^-$$

(This is effectively what happens if moisture gets into a flask containing EtMgBr.) Or from the other point of view, the tendency of OH^- to remove a proton from ethane is essentially *zero*.

To get examples of acid/base reactions that have an unfavourable position of equilibrium (i.e. that do not 'go'), all we have to do is consider all the reverse processes in Table 7.2. For example, an alkynide ion does not remove a proton from ammonia, the cyclopentadienyl ion does not remove a proton from triphenylmethane, hydroxide ion does not react with an alkane, and so on.

7.4 Solubility tests

Several simple qualitative tests for organic compounds depend on an acid/base reaction going virtually to completion and hence the organic compound dissolving in acid or base, *assuming of course that it is not simply soluble in water to start with*. For instance we can see from Table 7.1 that benzoic acid is a stronger acid than carbonic acid (H_2CO_3). Therefore benzoic acid will donate its proton to HCO_3^- (and H_2CO_3 then breaks down to CO_2 and H_2O). On the other hand, phenol is a weaker acid than carbonic acid and thus will not react with bicarbonate.

We can also see from the table that the simple neutral organic compounds that are the strongest bases are amines ($R{-}NH_2$, $Ar{-}NH_2$, guanidine, etc.). Only these are stronger bases than water and hence only these will react with a solution of the conjugate acid of water (H_3O^+). Thus of all classes of organic compound *only amines dissolve in dilute HCl* (with the proviso, of course, that many small organic compounds such as methanol, propanone, ethanal, etc. will also dissolve in dilute HCl simply because they are water soluble).

It is interesting to contrast EtOH and EtSH with respect to their reactivity towards NaOH solution. We can see from Table 7.2 that EtSH is a stronger acid than water which in turn is a stronger acid than EtOH. Thus, ethanol *dissolves* in dilute NaOH because it is soluble in water (why?), but it does not *react* with OH^-. On the other hand, ethanethiol (which is *not* very soluble in water) dissolves in dilute NaOH because it *does* react with OH^-.

7.5 Zwitterions and amino acids

The data in Table 7.1 show that in general the following reaction favours salt formation (when R′ is aliphatic):

$$\underset{\text{(stronger acid)}}{R{-}COOH} + \underset{\text{(stronger base)}}{R'{-}NH_2} \rightleftharpoons \underset{\text{(weaker base)}}{R{-}COO^-} + \underset{\text{(weaker acid)}}{R'{-}NH_3^+}$$

The same thing applies to aliphatic amino acids in which the $-COOH$ and $-NH_2$ groups are within the same molecule. Therefore, for example, glycine is $H_3\overset{+}{N}CH_2COO^-$ (and not H_2NCH_2COOH). (A dipolar ion such as glycine is called a **zwitterion**—literally a 'mongrel ion'.) We can also see from Table 7.1 that an *aromatic* amine is much less basic than an aliphatic one [for an explanation, see section 10.5(b)]; in fact, aniline is only very slightly more basic than the benzoate ion. Thus, the fact that 4-aminobenzoic acid is *not* a zwitterion is not really surprising. On the other hand a benzenesulphonic acid is appreciably more acidic than an aromatic carboxylic acid (see Table 7.1), acidic enough to protonate an aromatic amine and thus to account for the zwitterionic nature of 4-aminobenzenesulphonic acid:

$p\text{-}HOOC\text{-}C_6H_4\text{-}NH_2$ [*not* $p\text{-}{}^-OOC\text{-}C_6H_4\text{-}\overset{+}{N}H_3$]

m.p. 188 °C

$p\text{-}{}^-O_3S\text{-}C_6H_4\text{-}\overset{+}{N}H_3$ [*not* $p\text{-}HO_3S\text{-}C_6H_4\text{-}NH_2$]

decomposes at 280–300 °C, without melting (typical of an organic salt)

7.6 Unfavourable acid/base reactions that nevertheless lead to products

By this title I refer to reactions of the general kind:

$$\underset{\text{(weaker acid)}}{HA} + \underset{\text{(weaker base)}}{B^-} \rightleftharpoons \underset{\text{(stronger base)}}{A^-} + \underset{\text{(stronger acid)}}{HB} \rightarrow \text{products}$$

Here we have an acid/base reaction in which the position of equilibrium favours the starting materials but in which the equilibrium can be continually displaced over to the right by the occurrence of further steps in the reaction (leading eventually to a good yield of products). The last three entries in Table 7.2 exemplify this situation, as will be explained.

(a) The Reimer–Tiemann reaction

[See section 6.6.]

$$CHCl_3 + OH^- \rightleftharpoons H_2O + Cl_3C^- \rightarrow \text{further reaction}$$

The first step is unfavourable. (Chloroform certainly does not dissolve in NaOH solution in the same way as ArOH or RCOOH would.) But the small amount of Cl_3C^- ion is continuously removed by further reaction (to CCl_2 and then with phenoxide ion, etc.).

(b) The aldol condensation

The first step here [in the case of acetone, see section 6.5(h)] is:

$$CH_3COCH_3 + OH^- \rightleftharpoons CH_3COCH_2^- + H_2O \rightarrow \text{product}$$

Again this equilibrium does not lie very far to the right (i.e. at any given time there is only a very small amount of the enolate ion actually present). By continuously removing the product as it is formed, however, the reaction can be driven over to completion.

(c) The Claisen condensation

[See also sections 2.11, 5.5(p), 6.5(k).] Once again the first step is unfavourable (ethoxide ion is a weaker base than the enolate ion):

$$EtO^- + CH_3COOEt \rightleftharpoons EtOH + {}^-CH_2COOEt \rightarrow \text{further steps}$$

The last step is:

$$CH_3COCH_2COOEt + EtO^- \rightleftharpoons [CH_3COCHCOOEt]^- + EtOH$$

In this last step, the forward direction of reaction *is* favourable. (Ethoxide ion is a stronger base than $[CH_3COCHCOOEt]^-$ [see Table 7.1] since the latter is stabilised by delocalisation of the negative charge over *two* carbonyl groups.) Hence this step constitutes the driving force for the whole sequence of reactions. (At the completion of reaction the mixture is acidified to re-form the β-ketoester as product.) Thus we can do a Claisen condensation under the same conditions with RCH_2COOEt but not with $R_2CHCOOEt$—in the latter case there would be no central α-hydrogen for EtO^- to remove in a favourable last step:

$$R_2CH{-}CO{-}CR_2{-}COOEt + EtO^- \not\rightarrow$$

No reaction; therefore the unfavourable equilibrium of the first step is not pulled over to the right.

To get round this problem though, we can start by using an excess of a *stronger base* than EtO^- (such as Ph_3C^-). Then the final step is:

$$R_2CH{-}CO{-}CR_2{-}COOEt + Ph_3C^- \rightleftharpoons \left\{ \begin{array}{c} R_2C^-{-}COCR_2COOEt \\ \updownarrow \\ R_2C{=}\overset{O^-}{\overset{|}{C}}{-}CR_2COOEt \end{array} \right\} + Ph_3CH$$

Now *this* H is removed

Note that ethoxide ion is not basic enough to remove completely a hydrogen alpha to only one carbonyl group, whereas the triphenylmethide ion is so—and all this is easily rationalised in terms of the pK_a values of Table 7.1.

(d) Epoxide formation

Another example that falls into this general category is reaction of alcohols with OH^-. As already mentioned this reaction does not 'go' and so a water-insoluble alcohol does not dissolve in NaOH solution either. However, the position of equilibrium is such that enough alkoxide ion is present to enable certain reactions to occur, such as an internal nucleophilic substitution:

$$-\overset{OH}{\overset{|}{\underset{|}{C}}}-\underset{Br}{\underset{|}{\overset{|}{C}}}- + OH^- \rightleftharpoons -\overset{O^-}{\overset{|}{\underset{|}{C}}}-\underset{Br}{\underset{|}{\overset{|}{C}}}- + H_2O$$

(the occurrence of this step pulls the unfavourable equilibrium over to the right)

$$\downarrow$$

$$>C\overset{O}{\frown}C< + Br^-$$

(e) Amide hydrolysis

Finally consider the effect of acid on an amide. We can see from the relative pK_a values in Table 7.1 that the following reaction *does not proceed to completion*. (H_3O^+ and the protonated amide are of comparable acidity.)

$$R{-}\overset{O}{\overset{\|}{C}}{-}NH_2 + H_3O^+ \rightleftharpoons R{-}\overset{^+OH}{\overset{\|}{C}}{-}NH_2 + H_2O$$

Thus, (unlike an amine) an amide does not dissolve in dilute HCl. However, during the acid-catalysed hydrolysis of an amide [section 5.5(s)], there will be sufficient of the protonated form present for the reaction to proceed through this to products:

$$H_2\ddot{O} \curvearrowright \underset{R}{\overset{NH_2}{C}}{=}\overset{+}{O}H \rightarrow \text{etc.}$$

7.7 Elimination reactions

All nucleophiles are bases and vice versa since as we have noted above the definition of both classes involves the ability to donate an electron-pair—either to a proton (transferred from some acid) or to another atom, usually carbon, that is positively or partially positively charged.

$$\text{Base:} \curvearrowright H{-}A \qquad \text{Nucleophile:} \curvearrowright {}^{+}C{-}$$

or

$$\text{Nucleophile:} \curvearrowright {-}\overset{\delta+}{C}{-}$$

This means that in carrying out **substitution** reactions with alkyl halides for instance, we often get **elimination** occurring also, either as a minor side reaction or as the predominant process.

(a) The E1 mechanism

Recall first what happens in the S_N1 mechanism for the reaction between, say, *t*-butyl bromide (2-bromo-2-methylpropane) and ethanol:

$$CH_3{-}\underset{CH_3}{\overset{CH_3}{C}}{-}Br \rightleftharpoons CH_3{-}C^{+}(CH_3)_2 + Br^{-}$$

(2-bromo-2-methylpropane) (a 3° carbocation, so relatively stable)

Then nucleophilic attack of ethanol:

$$CH_3CH_2\ddot{O}H \curvearrowright (CH_3)_2C^{+}{-}CH_3 \longrightarrow CH_3CH_2{-}\underset{H}{\overset{+}{O}}{-}C(CH_3)_3 \xrightarrow{-H^{+}} \text{product (an ether)}$$

A competing reaction is with ethanol acting as a **base** rather than as a nucleophile:

$$CH_3CH_2\ddot{O}H \quad H{-}CH_2{-}\overset{+}{C}(CH_3)_2 \longrightarrow CH_3CH_2\overset{+}{O}H_2 + CH_2{=}C(CH_3)_2 \quad \text{(an alkene)}$$

This is of course simply an acid/base reaction on the Lowry–Bronsted definition. Ethanol (the base) accepts a proton; the *t*-butyl carbocation (the acid) donates a proton.

Generally speaking the product from the S_N1 process predominates over that from E1. If we desire the alkene as the major product we would use a stronger base in place of ethanol, but then the mechanism switches over to that known as E2.

(b) The E2 mechanism

If we treat *t*-butyl bromide with ethoxide ion (a relatively strong base) instead of ethanol, the mechanism of the elimination reaction that occurs is:

$$EtO^- \quad H{-}CH_2{-}C(CH_3)_2{-}Br \longrightarrow EtOH + CH_2{=}C(CH_3)_2 + Br^-$$

(Ethoxide ion does not wait for the alkyl halide to dissociate, but rather it attacks the alkyl halide directly.) This is a single-step concerted bimolecular process; the abstraction of the proton, the formation of the double bond, and the loss of the leaving group all happen at more or less the same time. The proton that is removed is extremely weakly acidic, about the same as in an alkane, and ethoxide ion of course cannot remove a proton from an alkane:

$$EtO^- \quad H{-}CH_3 \not\longrightarrow [EtOH + CH_3^-]$$

The reason ethoxide ion can remove the proton in the case of the alkyl halide is that at the same time we are producing the stable species $CH_2{=}C(CH_3)_2$ and Br^-, *not* the highly unstable, strongly basic, carbanion:

$$^-CH_2{-}C(CH_3)_2{-}Br$$

This E2 mechanism is strongly favoured for **tertiary** alkyl halides because (i) there is steric hindrance to the alternative substitution reaction, and (ii) the product alkene carries at least two alkyl groups and so is relatively stable.

We can now see why, for example, the reaction of sodium ethoxide with *t*-butyl bromide (2-bromo-2-methylpropane) is no good as an example of Williamson's ether synthesis [see section 5.3].

With **primary** alkyl halides (which are sterically unhindered) and a base-cum-nucleophile such as ethoxide ion, the S_N2 mechanism is favoured.

e.g.

$$EtO^- + H-\underset{CH_3}{\overset{H}{C}}-Br \xrightarrow{S_N2} Et-O-Et + Br^-$$

(major product)

$$\left[EtO^- + H-\underset{H}{\overset{H}{C}}-\underset{H}{\overset{H}{C}}-Br \xrightarrow{E2} EtOH + CH_2{=}CH_2 + Br^- \right]$$

(minor product)

With **secondary** alkyl halides, appreciable amounts of both elimination and substitution products are often formed. The proportion of E2 can be increased by raising the temperature and by using a strong bulky base such that substitution is sterically unfavourable.

e.g.

$$CH_3-\underset{CH_3}{\overset{CH_3}{C}}-O^- + H-CH_2-\overset{CH_3}{CH}-Br \xrightarrow{E2} (CH_3)_3C-OH + CH_2{=}CH-CH_3 + Br^-$$

(*t*-butoxide) (2-bromopropane) (major product)

On the other hand a weaker base favours the S_N2 mechanism, e.g.

$$CH_3CH_2S^- + \underset{CH_3}{\overset{CH_3}{CH}}-Br \xrightarrow{S_N2} CH_3CH_2-S-CH(CH_3)_2 + Br^-$$

(2-bromopropane) (major product)

[In broad terms a weak base is also often a weak nucleophile, and a strong base is often a powerful nucleophile, but there is by no means a faithful parallel between the two. For instance, in the last example, EtS^- is a weaker base than EtO^- but it is a better nucleophile. The reason why we should not expect nucleophilicity and basicity to go exactly hand-in-hand is that nucleophilicity is a measure of the **rate** at which a nucleophile attacks a compound, whereas basicity is not a measure of the rate at which a base accepts a proton—but rather a measure of the **position of equilibrium** between a base and its conjugate acid.]

8

Leaving Groups

8.1 Introduction

It is appropriate to consider leaving groups at this point, immediately after having looked at acid and base strength, since there is a strong correlation between the ability of a particular ion or molecule to act as a leaving group and the basicity of that ion or molecule. This was pointed out in Table 7.1; as base strength decreases, leaving group ability increases.

A strong base is a poor leaving group A weak base is a good leaving group

Consider what a leaving group is actually doing in its simplest terms—it is departing, with the bonding pair of electrons, from another atom (usually carbon):

$$C \!-\!\!\!\!\!\vdots\!-\! X$$

(Think of this process as it happens for instance in an S_N2 reaction, or in the breakdown of the tetrahedral intermediate after nucleophilic attack on an acyl compound.) Similarly, when an acid (HX) ionises to give a base (X^-) the 'leaving group' is this time departing from a proton:

$$H \!-\!\!\!\!\!\vdots\!-\! X$$

Intuitively it is not surprising then that there should be a relation between the ease of loss of X^- from R–X, and the ease of loss of X^- from H–X. For example, the usual order of reactivity of alkyl halides in S_N1 and S_N2 reactions is:

$$R\text{–}I > R\text{–}Br > R\text{–}Cl > R\text{–}F$$

and this is in accord with the strengths of the halogen acids:

$$H\text{–}I > H\text{–}Br > H\text{–}Cl > H\text{–}F$$

We should, however, appreciate the point that basicity is to do with the **position of equilibrium** between a base and its conjugate acid, whereas leaving group ability (like nucleophilicity, as mentioned at the end of the previous chapter) is to do with the **rate** at which a particular process occurs. We are therefore perhaps a little fortunate that there should be such a close and useful correlation between these different sorts of measurements.

In all the reactions met below you should be aware of the position of the leaving group in the order of basicity of Table 7.1. If it is a very weak base, it will be a good leaving group and the reaction will be relatively fast or easy. If it is a moderately strong base, it will be a poorer leaving group and then the reaction will be slower or more difficult to carry out. (Or perhaps the reaction can be changed in some way to render the leaving group better.) Finally, if the group is an extremely strong base, then it will be such a bad leaving group that it just will not leave.

8.2 Departure of a leaving group in the reactions of saturated compounds

In this section we will be considering all those reactions that can mechanistically be described as S_N1, S_N2, E1 and E2. All these have a leaving group, of course:

$$-\overset{|}{\underset{|}{C}}-LG \underset{}{\overset{S_N1}{\rightleftharpoons}} -\overset{|}{\underset{|}{C^+}} + LG^- \xrightarrow{Nuc} \text{product}$$

(Nuc. = Nucleophile)

$$-\overset{|}{\underset{|}{C}}-LG \overset{E1}{\rightleftharpoons} -\overset{|}{\underset{|}{C^+}} + LG^- \xrightarrow{-H^+} \text{product}$$

$$Nuc^- \; -\overset{|}{\underset{|}{C}}-LG \xrightarrow{S_N2} Nuc-\overset{|}{\underset{|}{C}}- + LG^-$$

$$Base^- \; H-\overset{|}{\underset{|}{C}}-\overset{|}{\underset{|}{C}}-LG \xrightarrow{E2} Base-H + \;>C=C<\; + LG^-$$

We are concentrating here on the leaving group. We have already considered how these various mechanisms are influenced by the nature of the compound, (e.g. 1°, 2°, or 3°), and the nature of the nucleophile or base.

(a) Leaving group is halide ion

Hydrogen halides are very strong acids. Halide ions are very weak bases. *Halide ions are very good leaving groups.* All these statements go together [see Table 7.1]. It is not surprising then that S_N1, S_N2, E1 and E2 reactions are most often typified by the reactions of alkyl halides, and that alkyl halides are such useful compounds in synthetic organic chemistry. The following are some examples.

$$(CH_3)_3C{-}Br \underset{}{\overset{S_N1}{\rightleftharpoons}} CH_3{-}\overset{+}{C}(CH_3)_2 + Br^-$$

$$CH_3{-}\overset{+}{C}(CH_3)_2 \xrightarrow{(i)\ H_2O,\ (ii)\ -H^+} (CH_3)_3C{-}OH \qquad Br^- \xrightarrow{Ag^+} AgBr$$

$$EtO^- + H{-}CH_2{-}CH(CH_3){-}Cl \xrightarrow{E2} EtOH + CH_2{=}CH{-}CH_3 + Cl^-$$

$$(EtOOC)(CH_3CO)CH^- + CH_3{-}I \xrightarrow{S_N2} (EtOOC)(CH_3CO)CH{-}CH_3 + I^-$$

(b) Leaving group is *p*-toluenesulphonate ('tosylate')

From Table 7.1 we see that an aromatic sulphonic acid is a strong acid and therefore the corresponding anion is a good leaving group. Tosylates are often used in substitutions and eliminations in a similar way to alkyl halides.

e.g.

$$(CH_3)_3N{:} + CH_3CH_2CH_2{-}O{-}SO_2{-}C_6H_4{-}CH_3 \xrightarrow{S_N2} (CH_3)_3\overset{+}{N}CH_2CH_2CH_3 + {}^-O_3S{-}C_6H_4{-}CH_3$$

(c) Leaving group is water

It is no use trying to prepare an alkyl halide from an alcohol and sodium bromide:

$$[Br^- \; R{-}OH \;\not\longrightarrow\; Br{-}R + OH^-]$$

The hydroxide ion is a fairly strong base and hence OH^- is a poor leaving group. Instead we have to do the reaction with HBr, or with NaBr/conc. H_2SO_4. Under these conditions an appreciable fraction of the alcohol is protonated and the leaving group is H_2O, not OH^-:

$$Br^- \; R{-}\overset{+}{O}H_2 \longrightarrow Br{-}R + H_2O$$

Water is a weaker base than OH^- (by a factor of about 10^{17}, see Table 7.1) and thus it is a much better leaving group. The reaction will now proceed (even though Br^- is not a very strong nucleophile).

A similar process is involved in ether formation from alcohols under acidic conditions. One alcohol molecule will not displace OH^- from another alcohol molecule:

$$[R{-}\ddot{O}H \; R{-}OH \;\not\longrightarrow\; R{-}\overset{+}{O}(H){-}R + OH^-]$$

But it will displace H_2O (a better leaving group) from a **protonated** alcohol molecule:

$$R{-}\ddot{O}H \; R{-}\overset{+}{O}H_2 \xrightarrow{S_N2} R{-}\overset{+}{O}(H){-}R + H_2O \xrightarrow{-H^+} \text{ether}$$

Under some conditions, (e.g. high temperature and high concentration of H_2SO_4), the protonated alcohol molecule does not wait around to be attacked by an unprotonated alcohol molecule, but instead loses a molecule of water and then a proton. (This is acid-catalysed dehydration of an alcohol.)

$$R{-}\overset{+}{O}H_2 \overset{E1}{\rightleftharpoons} R^+ + H_2O$$

$$R^+ \xrightarrow{-H^+} \text{alkene}$$

Again there is virtually no tendency for OH^- to act as a leaving group in this E1 process, only H_2O:

$$[R{-}OH \not\rightleftharpoons R^+ + OH^-]$$

Do not make the mistake of thinking that protonation will work for any intended substitution with an alcohol. For instance, the following reaction is not possible as OH^- is too poor a leaving group:

$$\left[H_3\ddot{N} \curvearrowright R{-}OH \longrightarrow H_3\overset{+}{N}{-}R + OH^- \longrightarrow H_2N{-}R + H_2O\right]$$

(1° amine)

But the reaction is *also* impossible under acidic conditions—we might imagine we would be improving the leaving group as in the examples above:

$$\left[H_3\ddot{N} \curvearrowright R{-}\overset{+}{O}H_2 \longrightarrow H_3\overset{+}{N}{-}R + H_2O\right]$$

However, as we can see from Table 7.1, ammonia is a stronger base than an alcohol and so acidification would protonate NH_3, not the ROH:

$$NH_3 + H^+ \rightarrow NH_4^+$$

—which means that our nucleophile (NH_3) has been effectively removed and so there can be no attack on ROH or $[ROH_2]^+$. (NH_4^+ of course is not a nucleophile. It does not have a lone pair of electrons.)

(d) Leaving group is ROH

The reaction of HBr or HI with an ether ('ether cleavage') is intimately 'connected' to the reaction of HBr or HI with an alcohol. In both cases halide ion is the nucleophile but it cannot displace the leaving group unless the latter is protonated. Thus:

$$[I^- \curvearrowright R{-}O{-}R' \not\longrightarrow I{-}R + {}^-O{-}R']$$

Instead:

$$I^- \curvearrowright R{-}\overset{+}{\underset{\substack{|\\H}}{O}}{-}R' \longrightarrow I{-}R + HO{-}R'$$

$$R{-}O{-}R' \xrightarrow{HI} R{-}\overset{+}{O}(H){-}R'$$

$$HO{-}R' \longrightarrow I{-}R'$$

[then further reaction as in section 8.2(c)]

Look up in Table 7.1 how much less basic ROH is than RO^-, and hence how much better a leaving group it is. Consider also that HI is a stronger acid than the protonated ether and so there will be a very substantial amount of the protonated ether for I^- to react with.

(e) Hydrolysis of epoxides

In the last example we saw that I^- will not attack an unprotonated ether. Neither will OH^-:

$$[HO^- \; R{-}O{-}R' \not\longrightarrow HO{-}R + {}^-O{-}R']$$

However this reaction *does* occur if we are dealing with the special case of a three membered cyclic ether (an epoxide):

$$HO^- + \text{epoxide} \longrightarrow HO{-}\overset{|}{\underset{|}{C}}{-}\overset{|}{\underset{|}{C}}{-}O^- \xrightarrow{H_2O} HO{-}\overset{|}{\underset{|}{C}}{-}\overset{|}{\underset{|}{C}}{-}OH$$

The leaving group is still an alkoxide ion (as it would be for an ordinary ether), it is just as strongly basic and hence would be expected to be as poor a leaving group. However, there is a new factor here that improves its ability to act as a leaving group—this is the **relief of ring-strain** that arises when the group leaves.

An alternative ring opening of epoxides can be carried out in acid conditions:

$$\text{epoxide} \underset{}{\overset{H^+}{\rightleftharpoons}} \text{protonated epoxide } (\overset{+}{O}H); \quad H_2O{:} + \text{protonated epoxide} \longrightarrow H_2\overset{+}{O}{-}\overset{|}{\underset{|}{C}}{-}\overset{|}{\underset{|}{C}}{-}OH \xrightarrow{-H^+} HO{-}\overset{|}{\underset{|}{C}}{-}\overset{|}{\underset{|}{C}}{-}OH$$

Even in dilute acid (where the protonation of the epoxide is not very extensive) the relief of ring-strain ensures that the reaction goes to completion. An ordinary ether of course is not susceptible to hydrolysis in dilute acid.

(f) Leaving group is trimethylamine

In the Hofmann elimination reaction a quaternary ammonium hydroxide is heated.

e.g.

$$CH_3CH_2CH_2CH_2\overset{+}{N}Me_3\ OH^- \rightarrow CH_3CH_2CH=CH_2 + H_2O + Me_3N$$

The mechanism is the familiar E2. We might imagine that a similar reaction could be brought about by treating, say, 1-butanamine with OH^-:

$$\left[HO^- \ H{-}\underset{\underset{CH_3}{|}}{\underset{CH_2}{|}}{C}{-}\underset{|}{C}{-}NH_2 \not\longrightarrow H_2O + \underset{\underset{CH_3}{|}}{\underset{CH_2}{|}}{CH}{=}CH_2 + NH_2^- \right]$$

By now, though, you should be able to grasp that NH_2^- is far too strong a base (too poor a leaving group) for this reaction to work. If the leaving group were NH_3 (instead of NH_2^-) then theoretically the reaction should be more successful, in the same way as H_2O is a better leaving group than OH^-, so we might propose:

$$\left[HO^- \ H{-}\underset{\underset{CH_3}{|}}{\underset{CH_2}{|}}{C}{-}\underset{|}{C}{-}\overset{+}{N}H_3 \longrightarrow H_2O + \underset{\underset{CH_3}{|}}{\underset{CH_2}{|}}{CH}{=}CH_2 + NH_3 \right]$$

But again this is not possible in practice. Instead we would merely get the following simple acid/base reaction:

$$\underset{\text{(stronger base)}}{HO^-} + \underset{\text{(stronger acid)}}{CH_3CH_2CH_2CH_2\overset{+}{N}H_3} \rightleftharpoons \underset{\text{(weaker acid)}}{H_2O} + \underset{\text{(weaker base)}}{CH_3CH_2CH_2CH_2NH_2}$$

This is why, if we want to eliminate nitrogen from an amine such as 1-butanamine, we first convert it to the **quaternary** hydroxide:

$$CH_3CH_2CH_2CH_2NH_2 \xrightarrow[\text{(S}_N\text{2, 3 times)}]{3CH_3I} CH_3CH_2CH_2CH_2\overset{+}{N}(CH_3)_3\ I^-$$

$$\downarrow \text{'AgOH' (moist silver oxide)}$$

$$CH_3CH_2CH_2CH_2\overset{+}{N}(CH_3)_3\ OH^- \quad + AgI$$

The 4° hydroxide is a fully dissociated ionic compound. There is no hydrogen on the nitrogen atom for OH^- to remove in a simple acid/base reaction as there was with $CH_3CH_2CH_2CH_2NH_3^+$. Now, therefore, there is no bar to a successful elimination reaction:

$$HO^- \; H{-}\underset{\underset{CH_3}{|}}{\underset{CH_2}{|}}{C}{-}\underset{|}{C}{-}\overset{+}{N}Me_3 \longrightarrow H_2O + \underset{\underset{CH_3}{|}}{\underset{CH_2}{|}}{CH}{=}CH_2 + NMe_3$$

(Trimethylamine is similar to ammonia in basicity and hence from Table 7.1 we would expect it to have moderate leaving group ability.)

8.3 Departure of a leaving group in the reactions of acyl compounds

There are many examples of reactions involving nucleophilic substitution or displacement at acyl carbon. As we have seen [section 5.5] these reactions involve two main steps—nucleophilic addition to give a tetrahedral intermediate followed by elimination of a leaving group:

$$Nuc^- \; \underset{\underset{R}{|}}{\overset{\overset{LG}{|}}{C}}{=}O \rightleftharpoons Nuc{-}\underset{\underset{R}{|}}{\overset{\overset{LG}{|}}{C}}{-}O^- \longrightarrow Nuc{-}\underset{\underset{R}{|}}{C}{=}O + LG^-$$

(a) Leaving group is halide ion

Acyl halides are very reactive towards nucleophilic attack partly because the carbonyl carbon has a substantial $\delta+$ charge [section 9.5(a)] and partly because the halide ion is such a good leaving group. Thus as we have seen, nucleophiles that will readily react with acyl halides include water, alcohols, phenols, ammonia and amines. Just as *alkyl* halides are so useful for alkylating other compounds, so *acyl* halides are invaluable for acylating compounds (e.g. turning alcohols and phenols into esters, and ammonia and amines into amides). [For examples, see sections 5.5(a), (b), (c), (d), (e), (f), (g), (h).]

(b) Leaving group is a carboxylate ion

A carboxylate ion is a stronger base than a halide ion (see Table 7.1) and so it is a poorer leaving group. This is reflected in one of the main ways of preparation of anhydrides:

$$R{-}COO^- + Cl{-}\overset{|}{C}(R'){=}O \longrightarrow R{-}\overset{O}{\overset{\|}{C}}{-}O{-}\overset{Cl}{\underset{R'}{C}}{-}O^- \longrightarrow R{-}\overset{O}{\overset{\|}{C}}{-}O{-}\underset{R'}{C}{=}O + Cl^-$$

The point here is that the intermediate *could* break down instead to expel $RCOO^-$ (which is merely the reverse of the first step):

$$R{-}\overset{O}{\overset{\|}{C}}{-}O{-}\overset{Cl}{\underset{R'}{C}}{-}O^- \longrightarrow R{-}\overset{O}{\overset{\|}{C}}{-}O^- + \overset{Cl}{\underset{R'}{C}}{=}O$$

However, since $RCOO^-$ is a poorer leaving group than Cl^-, the intermediate prefers to expel the latter and a good yield of the anhydride is obtained.

An anhydride can be used for acylating compounds (alcohols, phenols, amines) in just the same way as an acyl halide. Since the leaving group is not so good, however, the reactions are rather slower and less vigorous. [For examples, see section 5.5(i).]

(c) Leaving group is RO^- or ROH

From Table 7.1 we can see that an alkoxide ion is much more basic than $RCOO^-$ or Cl^-, and therefore is a much poorer leaving group. Thus, an **ester** is much less reactive towards water, for example, than is an acyl halide or anhydride. However, it can be hydrolysed by sodium hydroxide solution:

$$HO^- + \overset{OR'}{\underset{R}{C}}{=}O \rightleftharpoons HO{-}\overset{OR'}{\underset{R}{C}}{-}O^- \longrightarrow HO{-}\underset{R}{C}{=}O + R'O^-$$

$$HO{-}\underset{R}{C}{=}O \xrightarrow{-H^+} RCOO^- \qquad R'O^- \xrightarrow{+H^+} R'OH$$

The driving force here is the formation, in the last step, of the resonance-stabilised carboxylate anion [see section 10.2(a)]. This factor outweighs the fact that $R'O^-$ is a rather poor leaving group.

If the attacking nucleophile is very powerful then alkoxide ion can be expelled quite readily. This occurs in the reaction of esters with Grignard reagents and with lithium aluminium hydride.

e.g.

$$CH_3^- \ (CH_3MgBr) + C_6H_5-C(=O)-OCH_2CH_3 \longrightarrow CH_3-C(C_6H_5)(OCH_2CH_3)-O^- \longrightarrow CH_3-C(=O)-C_6H_5 + CH_3CH_2O^-$$

further reaction → 3° alcohol

$$H^- \ (LiAlH_4) + (CH_2)_{16}CH_3-C(=O)-OCH_3 \longrightarrow H-C((CH_2)_{16}CH_3)(OCH_3)-O^- \longrightarrow H-C(=O)-(CH_2)_{16}CH_3 + CH_3O^-$$

further reduction → 1° alcohol

An ester can be made more reactive to a weaker nucleophile such as **water** by using acidic conditions:

$$RCOOR' \overset{H^+}{\rightleftharpoons} R-C(OR')=\overset{+}{O}H \overset{H_2O:}{\rightleftharpoons} H_2\overset{+}{O}-C(R)(OR')-OH \overset{\pm H^+}{\rightleftharpoons} HO-C(R)(H\overset{+}{O}R')-\ddot{O}H \rightleftharpoons HO-C(R)=\overset{+}{O}H + R'OH \overset{-H^+}{\rightleftharpoons} RCOOH$$

The nucleophile (H_2O) attacks the protonated ester more readily than the ester itself (although since the ester is so very weakly basic [see Table 7.1] there will be very little protonated ester present at any given time). The other way in which acid assists the reaction is by ensuring that the leaving group is R′OH instead of the much more strongly basic $R'O^-$. 'Connect' this with ether cleavage [section 8.2(d)] where again a molecule of alcohol is the leaving group rather than an alkoxide ion.

(d) Acid-catalysed esterification

e.g.

$$CH_3COOH \xrightarrow{H^+} CH_3\overset{+}{C}(OH)_2 \;\;(\text{i.e. } CH_3C(OH){=}\overset{+}{O}H)$$

$$CH_3\ddot{O}H + CH_3C(OH){=}\overset{+}{O}H \rightleftharpoons CH_3{-}\overset{+}{O}(H){-}C(OH)_2{-}CH_3 \underset{}{\overset{\pm H^+}{\rightleftharpoons}} CH_3{-}O{-}C(\overset{+}{O}H_2)(\ddot{O}H){-}CH_3$$

$$\rightleftharpoons CH_3{-}O{-}C(CH_3){=}\overset{+}{O}H + H_2O \underset{}{\overset{-H^+}{\rightleftharpoons}} CH_3COOCH_3$$

$$CH_3\ddot{O}H \overset{H^+}{\rightleftharpoons} CH_3\overset{+}{O}H_2$$

Note:

(i) This mechanism is exactly the same as the reverse of acid-catalysed ester hydrolysis in the previous section.

(ii) Acid (e.g. H_2SO_4) *helps* the reaction by protonating the ester and making it more electrophilic (that is, more attractive to the nucleophile, CH_3OH), and by ensuring that the leaving group is H_2O (a good one), not OH^- (a poor one).

(iii) Acid *hinders* the reaction by tending to protonate the alcohol, which destroys its nucleophilicity. Thus the overall rate is a balance between factors that favour and disfavour the reaction. (In this case the reaction proceeds in the forward direction if an excess of either the alcohol or the carboxylic acid is used.) If the nucleophile is much more basic than CH_3OH (e.g. NH_3) then it cannot attack the carboxylic acid nucleophilically *with or without* acid catalysis. If H_2SO_4 is added it completely protonates the nucleophile, of course:

$$NH_3 + H^+ \rightarrow NH_4^+$$

and if H_2SO_4 is not added the carboxylic acid itself is acidic enough to protonate ammonia or an amine. This reaction does **not** occur,

$$\left[H_3\ddot{N} \curvearrowright \underset{R}{\overset{OH}{C}}{=}O \;\not\longrightarrow\; H_3\overset{+}{N}-\underset{R}{\overset{OH}{C}}-O^- \right]$$

but instead this reaction occurs

$$H_3\ddot{N} \curvearrowright H-\ddot{O}-\underset{R}{C}{=}O \longrightarrow NH_4^+ + \underset{R}{COO^-}$$

(e) Leaving group is ammonia or an amine

The acid-catalysed hydrolysis of an amide is essentially identical in mechanism to the acid-catalysed hydrolysis of an ester. Again acid promotes the reaction by making the amide more electrophilic to the nucleophile (H_2O) and by improving the leaving group (which is NH_3 instead of NH_2^-, $R{-}NH_2$ instead of $R{-}NH^-$, or R_2NH instead of R_2N^-).

e.g.

$$CH_3CONHCH_3 \overset{H^+}{\rightleftharpoons} H_2\ddot{O} \curvearrowright \underset{CH_3}{\overset{NHCH_3}{C}}{=}\overset{+}{O}H \rightleftharpoons H_2\overset{+}{O}-\underset{CH_3}{\overset{NHCH_3}{C}}-OH \overset{\pm H^+}{\rightleftharpoons} HO-\underset{CH_3}{\overset{\overset{+}{N}H_2CH_3}{C}}-\ddot{O}H$$

$$\longrightarrow HO-\underset{CH_3}{C}{=}\overset{+}{O}H + CH_3NH_2$$

$$HO-\underset{CH_3}{C}{=}\overset{+}{O}H \xrightarrow{-H^+} CH_3COOH \qquad CH_3NH_2 \xrightarrow{+H^+} CH_3\overset{+}{N}H_3$$

$$CH_3COOH + CH_3\overset{+}{N}H_3$$

Alkaline hydrolysis of an amide in its simplest terms can be thought of as being similar to alkaline hydrolysis of an ester [section 8.3(c)]:

$$HO^- + R{-}C(NH_2){=}O \rightleftharpoons HO{-}C(NH_2)(R){-}O^- \longrightarrow HO{-}C(R){=}O + NH_2^-$$

$$HO{-}C(R){=}O \xrightarrow{-H^+} RCOO^- \qquad NH_2^- \xrightarrow{+H^+} NH_3$$

However, the amide ion (NH_2^-) is such a **very strong base** that it is an **extremely poor leaving group** (see Table 7.2). Thus, as already mentioned [section 5.5(t)], it is perhaps better to think of NH_2^- never actually being expelled as such, but being simultaneously protonated by a molecule of water, thus:

$$H{-}OH + HO{-}C(NH_2)(R){-}O^- \longrightarrow HO{-}C(R){=}O + NH_3 + OH^-$$

$$HO{-}C(R){=}O \longrightarrow RCOO^-$$

8.4 Leaving groups in the reactions of aldehydes and ketones

After nucleophilic addition to an *aldehyde or ketone* in general we do *not* get expulsion of a leaving group and breakdown of the tetrahedral intermediate species back to an acyl compound.

e.g.

$$Nuc^- + C_6H_5{-}C(H){=}O \rightleftharpoons Nuc{-}C(H)(C_6H_5){-}O^- \quad [\not\longrightarrow Nuc{-}C(C_6H_5){=}O + H^-]$$

$$Nuc^- + (CH_3)_2C{=}O \rightleftharpoons Nuc{-}C(CH_3)_2{-}O^- \quad [\not\longrightarrow Nuc{-}C(CH_3){=}O + CH_3^-]$$

This means that there are many examples of nucleophilic addition to aldehydes and ketones where the product contains tetrahedrally coordinated carbon—for example, hydrates, alcohols (after attack of $LiAlH_4$ or RMgBr), cyanohydrins, aldols, hemiacetals, etc. [See section 5.6.] The reason here is of course that R^- and H^- are such extremely strong bases that they will not act as leaving groups (see Table 7.1). As we have seen, one minor exception to this general rule is the Cannizzaro reaction [section 5.6(k)]; even here we should not think of H^- leaving in the same way as Cl^- would:

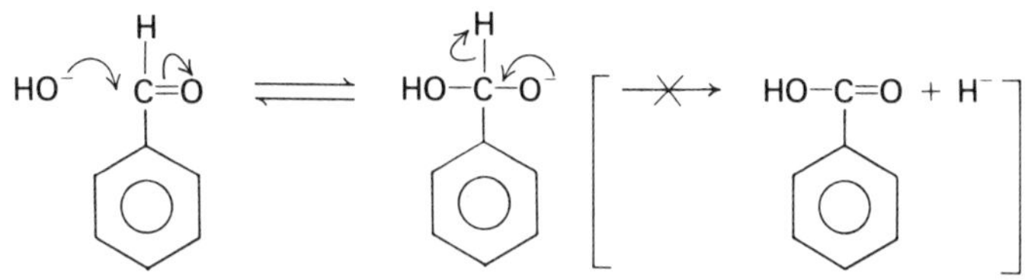

HO⁻ + PhCH=O ⇌ HO–CH(Ph)–O⁻ [—✕→ HO–C(Ph)=O + H^-]

but as being directly transferred to the carbonyl group of a second aldehyde molecule without ever becoming free:

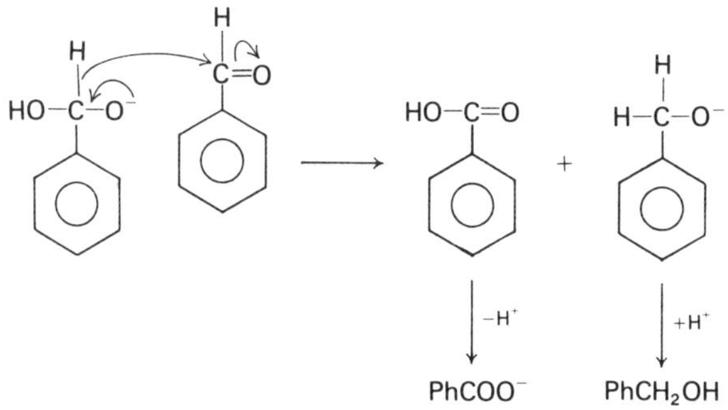

HO–CH(Ph)–O⁻ + PhCH=O ⟶ HO–C(Ph)=O + H–CH(Ph)–O⁻

HO–C(Ph)=O —($-H^+$)→ $PhCOO^-$

H–CH(Ph)–O⁻ —($+H^+$)→ $PhCH_2OH$

(a) Leaving group is H_2O after nucleophilic addition to aldehydes and ketones

Consider for example a hemiacetal:

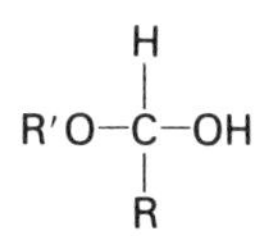

R′O–CH(R)–OH

Neither the R nor the H can act as a leaving group, but the OH *can* if it is first protonated such that it leaves as a molecule of water. (What other examples have we seen above where OH^- does not leave but H_2O does?)

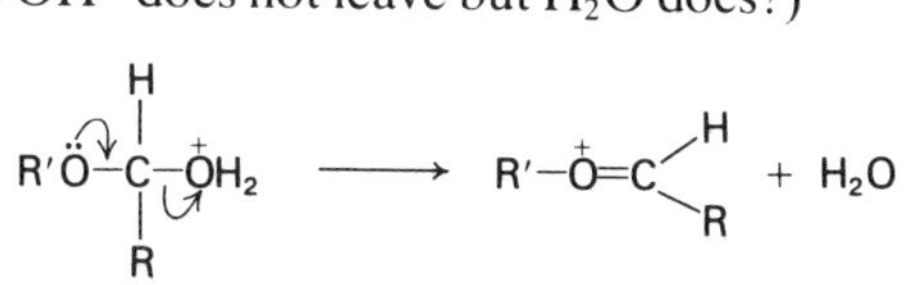

$$R'\ddot{O}\text{–}CH(R)\text{–}\overset{+}{O}H_2 \longrightarrow R'\text{–}\overset{+}{O}\text{=}C(H)R + H_2O$$

Then a second molecule of alcohol can nucleophilically attack this positively charged species giving eventually an *acetal*. [Try following the mechanism through before checking back to section 5.6(i).]

The same situation arises in the reaction of an aldehyde or ketone with, for example, 2,4-dinitrophenylhydrazine:

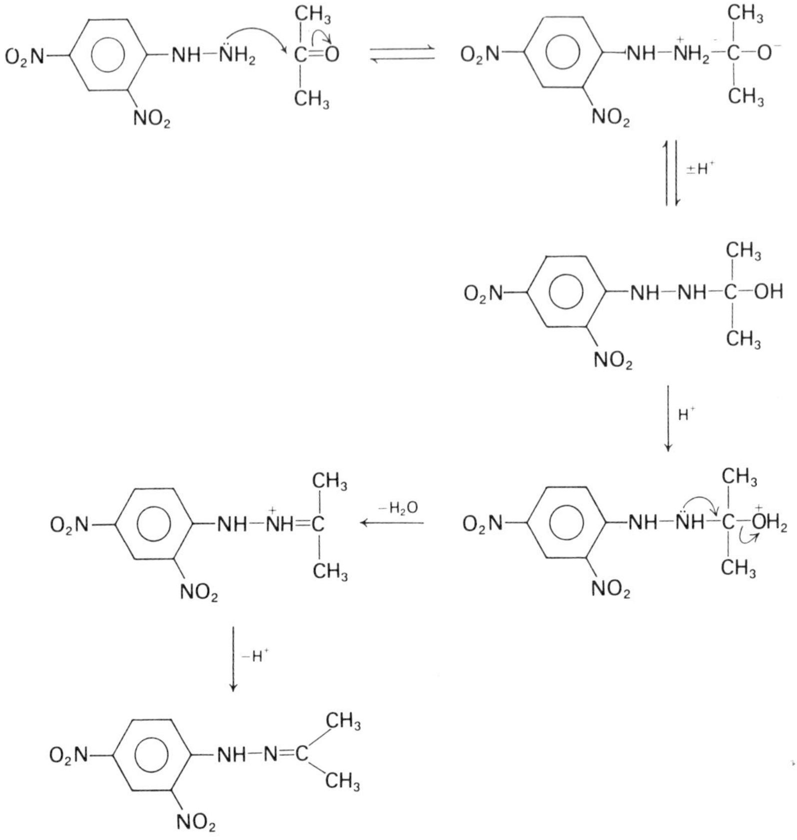

Notice how similar the step involving loss of H_2O is to the breakdown of the protonated hemiacetal above.

(b) Leaving group is Cl_3C^-

When acetone (propanone) is chlorinated in alkaline conditions [see section 4.6], the reaction proceeds to give CH_3COCCl_3 which is then attacked nucleophilically by OH^-:

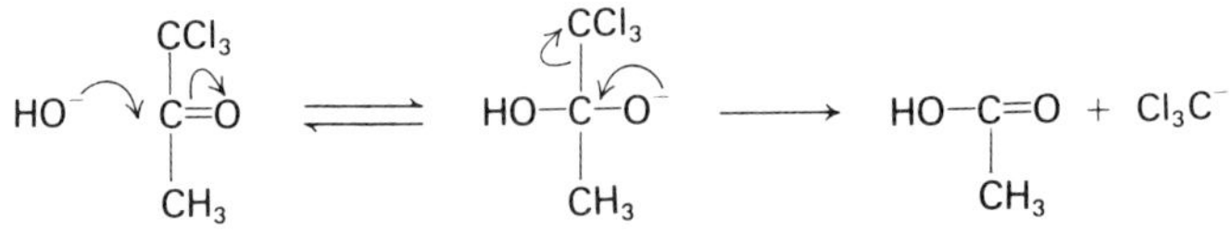

Note that a similar reaction does not happen with acetone itself:

$$HO^- + (CH_3)_2C{=}O \rightleftharpoons HO{-}C(CH_3)_2{-}O^- \quad \left[\not\longrightarrow HO{-}C(CH_3){=}O + CH_3^- \right]$$

That is, Cl_3C^- can act as a leaving group but CH_3^- cannot. Why is this? The explanation, as for all these reactions involving leaving groups, is in terms of relative base strengths. From Table 7.1 we can see that $CHCl_3$ is considerably more acidic than a simple alkane. In other words, Cl_3C^- is much less basic than CH_3^-, and hence the former can act as a leaving group in the reaction above whereas the latter cannot. It is also clear from Table 7.1 that Cl_3C^- is nevertheless still quite a strong base (relative to OH^- for instance) and thus as soon as it is expelled it undergoes the following acid/base reaction:

$$\underset{\text{(stronger base)}}{CCl_3^-} + \underset{\text{(stronger acid)}}{H_2O} \rightleftharpoons \underset{\text{(weaker acid)}}{CHCl_3} + \underset{\text{(weaker base)}}{OH^-}$$

Similar explanations are involved in the production of $CHBr_3$ and CHI_3 in analogous reactions (such as the 'iodoform test').

9

Inductive Effects

9.1 Introduction

In a simple symmetrical bond such as the C—C bond in ethane, the electrons are of course shared *equally* between the two atoms. However, in the majority of bonds this is not the case. For example, chlorine is a more electronegative element than carbon; this means that, in chloromethane, for example, the bonding pair of electrons is drawn somewhat closer to chlorine than to carbon:

$$\overset{\delta+}{CH_3}—\overset{\delta-}{Cl}$$

(The distribution of charge is reflected in the sizeable dipole moment of 1.94 D.) We often describe this as the inductive effect of chlorine and represent it by an arrow on the bond, thus:

$$CH_3 \rightarrow Cl$$

Chlorine has an **electron-withdrawing** inductive effect. Most other substituents also have an inductive effect in the same direction; the main exception is found in alkyl groups which have an **electron-donating** or 'electron-pushing' inductive effect that we can represent as, for example:

$$C_6H_5 \leftarrow CH_3$$

Why study inductive effects? Because as we will see, in many cases the course of a reaction or the stability of a compound (and hence its reactivity or lack of it) can be rationalised in terms of inductive effects. In general, we are concerned with whether a substituent 'pulls' or 'pushes' electrons with respect to some other part of the molecule that is of positive or negative nature. (This does not necessarily mean a **full** positive or negative charge.) Four possible cases can arise:

1.	$X \rightarrow Y(-)$	destabilising
2.	$X \rightarrow Y(+)$	stabilising
3.	$X \leftarrow Y(-)$	stabilising
4.	$X \leftarrow Y(+)$	destabilising

In category (1), X has an electron-donating inductive effect (e.g. CH_3); it tends to push electrons to another part of the molecule (Y) that is already of high electron density. This is clearly an unfavourable situation ('like charges repel') and has a *de*-stabilising influence on the species. On the other hand, if X tends to donate negative charge towards a *positive* centre [as in category (2)], then the species is stabilised. In a similar way the situations represented in (3) and (4) have stabilising and destabilising influences respectively.

We will now look at several superficially unconnected aspects of organic chemistry and see how they all fall neatly into categories involving the various ways in which inductive effects can operate.

9.2 Inductive effect of alkyl groups on negative centres

(a) Acidity of formic and acetic acids

Acetic (ethanoic) acid has $pK_a = 4.8$ and formic (methanoic) acid has $pK_a = 3.7$. This means that acetic acid is weaker than formic acid, it loses its proton less readily than does formic acid, which in turn means that the acetate ion is *less stable* relative to acetic acid than the formate ion is relative to formic acid. We can see in terms of our process (1) above, X⤚Y (−), that the methyl substituent in the acetate ion is tending to push electrons towards the negatively charged carboxylate group [CH_3⤚COO^- (a destabilising inductive effect)]. No such destabilising influence is present in the formate ion (H—COO^-).

(b) Course of the haloform reaction

When a methyl ketone such as RCH_2COCH_3 is halogenated in basic solution, the first product (not isolated) is RCH_2COCH_2X, not $RCHXCOCH_3$. Why is this? One reason is that the methyl group is the more sterically accessible part of the molecule. Another reason is as follows. The mechanism involves formation of an enolate ion that is then electrophilically attacked by halogen [see section 4.6]—the question therefore comes down to which of the following enolate ions is preferred?

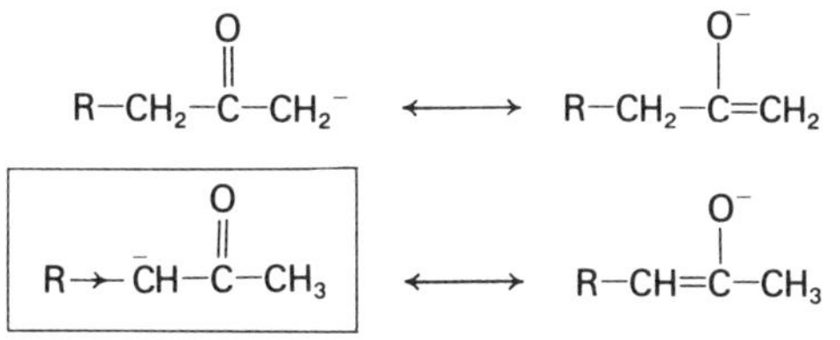

$$R{-}CH_2{-}\overset{\overset{\Large O}{\|}}{C}{-}CH_2^- \longleftrightarrow R{-}CH_2{-}\overset{\overset{\Large O^-}{|}}{C}{=}CH_2$$

$$\boxed{R \rightarrowtail \overset{-}{C}H{-}\overset{\overset{\Large O}{\|}}{C}{-}CH_3} \longleftrightarrow R{-}CH{=}\overset{\overset{\Large O^-}{|}}{C}{-}CH_3$$

Answer: the *first* one is, because we can see that in one resonance form of the second, the inductive effect of the alkyl substituent destabilises the neighbouring negative charge. Thus we can see why, in the haloform reaction, substitution starts in the *methyl* group, not in whatever group is on the other side of the carbonyl group. (We will see later [section 9.4(a)] why substitution continues in this group to give $-CX_3$.)

9.3 Inductive effect of alkyl groups on positive centres

(a) Relative stability of carbocations

The classic example in which the inductive effect of alkyl groups is invoked is in the explanation of the order of stability of carbocations:

$$Me \rightarrow \overset{+}{C}(\leftarrow Me)_2 > Me \rightarrow \overset{+}{C}H(\leftarrow Me) > Me \rightarrow \overset{+}{C}H_2 > H-\overset{+}{C}H_2$$

Clearly, we are now into category (2), X→Y(+). The substituent tends to donate electrons towards the positive centre, thereby stabilising it. The more alkyl substituents there are, the more stable is the carbocation. This order of stability is reflected in various experimental observations:

(i) It can be shown that *tertiary* alkyl halides react with nucleophiles according to the S_N1 mechanism (involving the intermediacy of a carbocation [sections 1.2(a), 5.3]).

(ii) *Tertiary* alcohols are dehydrated under much milder conditions than 2° or 1° ones (because the 3° carbocation is easier to form).

(iii) When reagents of the form H–X add to unsymmetrical alkenes such as $CH_3CH{=}CH_2$ or $(CH_3)_2C{=}CH_2$, the orientation of addition follows Markovnikov's rule and can be rationalised in terms of the relative stability of the possible carbocations that are involved in the reaction mechanism [section 1.3].

(iv) In reactions involving the generation of a carbocation, this species may rearrange to give a more stable carbocation, (e.g. 1°→2°, 1°→3°, or 2°→3°), before the final products are obtained [section 1.4]. This frequently happens for instance in the dehydration of alcohols,

e.g.

$$CH_3CH_2CH_2CH(CH_3)CH_2OH \xrightarrow{H_2SO_4} CH_3CH_2CH_2C(CH_3){=}CH_2 + CH_3CH_2CH{=}C(CH_3)-CH_3$$

and in Friedel–Crafts alkylation reaction,

e.g.

$$C_6H_6 + (CH_3)_2CH-CH_2Br \xrightarrow{AlBr_3} C_6H_5-C(CH_3)_3 + HBr$$

(Check that you can recall the *details* of these mechanisms.)

[The relative stability of carbocations is further addressed in section 10.10(a).]

(b) Alkyl groups activate the benzene ring towards electrophilic substitution

As we will see again shortly in Chapter 10, electrophilic aromatic substitution involves an intermediate cation that may or may not be stabilised by the presence of various substituents. If the substituent is an alkyl group then the cation is **stabilised by the inductive effect** [again an example of our general category (2): X→Y(+), see section 9.1] and therefore the substitution reaction proceeds faster (or under milder conditions) than with benzene.

e.g.

CH_3 (toluene) $\xrightarrow[H_2SO_4]{HNO_3}$ (cation with CH_3, + , H, NO_2) $\xrightarrow{-H^+}$ CH_3/NO_2 (para product) (+ the ortho isomer)

Toluene is nitrated 25 times faster than benzene. [We will consider why the product is mainly ortho and para, not meta, in section 10.7(b).]

The same point is behind the frequent occurrence of **poly**substitution in Friedel–Crafts alkylation. The first formed product is more reactive than benzene towards electrophiles (because of the inductive effect of the alkyl substituent) and hence tends to be substituted again.

e.g.

benzene $\xrightarrow[AlCl_3]{EtCl}$ Et-benzene + [Et/Et (para), Et/Et (ortho), Et/Et/Et , etc.]

(c) Rate of addition of bromine to alkenes

Just as alkyl groups can activate a benzene ring towards electrophiles, so can they activate an alkene towards electrophiles.

e.g.

$Me_2C{=}CMe_2 + Br_2 \longrightarrow$ (bromonium ion: Me_2C—CMe_2 bridged by Br^+) $\xrightarrow{Br^-} Me_2CBr{-}CBrMe_2$

The four methyl groups' electron-pushing inductive effects all help to stabilise the intermediate positively charged bromonium ion. The result is that this alkene adds bromine 13 times faster than does ethene. Note that this factor evidently outweighs any steric hindrance on the part of the four methyl groups to the approach of bromine.

[Another factor that may help to explain why the alkene above (2,3-dimethyl-2-butene) reacts faster than ethene is the relief of steric compression of the four alkyl groups; in the starting material they are co-planar (that is, eclipsed), but in the product they can take up a staggered conformation.]

(d) Relative basicity of amines

Consider these pK_a values:

NH_4^+	9.25
$CH_3NH_3^+$	9.64

$CH_3NH_3^+$ is a slightly weaker acid than NH_4^+ or in other words, CH_3NH_2 (methanamine) is a slightly stronger base than NH_3. It is sometimes said that this is because $CH_3NH_3^+$ is somewhat stabilised by the inductive effect of the methyl group and thus is slightly easier to form from CH_3NH_2 than NH_4^+ is from NH_3:

$$CH_3 \rightarrowtail N^+H_3 \quad \text{versus} \quad H-N^+H_3$$

[Again we have an example from category (2): X⤚Y (+), a stabilising influence.]

However, this is not a very impressive example of the inductive effect since the pK_a difference is only small, and the trend breaks down when the series is continued to $(CH_3)_2NH_2^+$ (pK_a = 9.72) and $(CH_3)_3NH^+$ (pK_a = 9.70). These various species and their corresponding free amines are stabilised to different extents by **solvation** with water molecules. Consequently any stabilisation due to the inductive effect of the methyl groups tends to be obscured.

9.4 Inductive effect of halogen on negative centres

(a) The course of the haloform reaction

We discussed in section 9.2(b) the reasons to expect R–CH_2–CO–CH_2X to be formed in preference to R–CHX–CO–CH_3 (from R–CH_2–CO–CH_3 and X_2/NaOH). The next problem is to find out why the reaction goes on to give R–CH_2–CO–CHX_2 (and then R–CH_2–CO–CX_3) rather than R–CHX–CO–CH_2X. Again we think in terms of the enolate ion that is involved in the mechanism. Which of the following alternatives will be preferred?

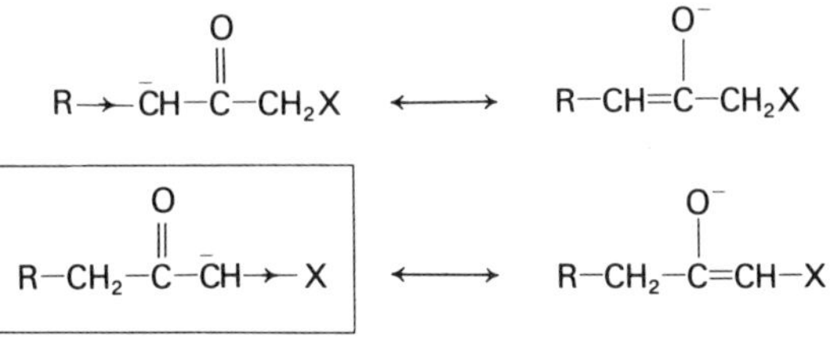

$$R\!\rightarrow\!\overset{-}{C}H-\overset{\overset{O}{\|}}{C}-CH_2X \longleftrightarrow R-CH=\overset{\overset{O^-}{|}}{C}-CH_2X$$

$$\boxed{R-CH_2-\overset{\overset{O}{\|}}{C}-\overset{-}{C}H\!\rightarrow\!X} \longleftrightarrow R-CH_2-\overset{\overset{O^-}{|}}{C}=CH-X$$

Answer: The *second*, because one of its resonance forms has the 'electron-pulling' inductive effect of halogen helping to stabilise the neighbouring negative charge. [This belongs to category (3): X←Y(−) see section 9.1.] In addition, as before the other possible enolate ion is *de*stabilised by the inductive effect of the R group. The upshot is that a methyl ketone gets systematically substituted in its methyl group to give $-CCl_3$, $-CBr_3$, or $-CI_3$ (even at the expense of introducing three quite bulky atoms at the same carbon).

The final step in the haloform reaction—cleavage by alkali—should now be revised. [See sections 5.5(v), 8.4(c).]

(b) Acidity of 1,1,1-trifluoropentane-2,4-dione

The pK_a of this compound ($CF_3COCH_2COCH_3$) is 4.7. In other words it is as acidic as acetic acid. (It would displace carbon dioxide from $NaHCO_3$ solution.) The pK_a of the fluorine-free analogue, acetylacetone, or pentane-2,4-dione is 9.0. (It is only as acidic as phenol.) The acidity of **both** compounds is associated with resonance-stabilisation of the corresponding anion [see section 10.3(e)], but only the fluorinated anion is also *further* stabilised by the powerful inductive effect of the $-CF_3$ group:

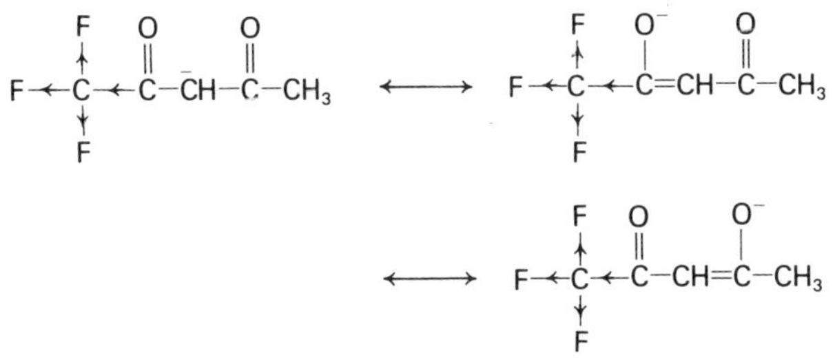

$$F\!\leftarrow\!\underset{\underset{F}{\downarrow}}{\overset{\overset{F}{\uparrow}}{C}}\!\leftarrow\!\overset{\overset{O}{\|}}{C}-\overset{-}{C}H-\overset{\overset{O}{\|}}{C}-CH_3 \longleftrightarrow F\!\leftarrow\!\underset{\underset{F}{\downarrow}}{\overset{\overset{F}{\uparrow}}{C}}\!\leftarrow\!\overset{\overset{O^-}{|}}{C}=CH-\overset{\overset{O}{\|}}{C}-CH_3$$

$$\longleftrightarrow F\!\leftarrow\!\underset{\underset{F}{\downarrow}}{\overset{\overset{F}{\uparrow}}{C}}\!\leftarrow\!\overset{\overset{O}{\|}}{C}-CH=\overset{\overset{O^-}{|}}{C}-CH_3$$

[This falls into category (3): X←Y(−), a stabilising influence.] This is just one example of how acidity is increased by electron-withdrawing substituents; the following section contains some more familiar examples.

(c) Relative acidity of halogen-containing acids

Consider these pK_a values:

CH_3COOH	4.76
$CH_2ClCOOH$	2.87
$CHCl_2COOH$	1.26
CCl_3COOH	0.63

Chloroacetic (chloroethanoic) acid is appreciably *stronger* than acetic acid; this means that it is relatively easy to form the CH_2ClCOO^- ion, and the reason is that the negative charge is stabilised by the inductive effect of the chlorine substituent:

$$Cl \leftarrow CH_2COO^- \quad \text{versus} \quad CH_3COO^-$$

[a stabilising influence—see category (3) in section 9.1].

The more chlorine substituents there are, the more strongly is the carboxylate anion stabilised. Trichloroacetic acid is more acidic than, for instance, H_3PO_4.

Consider also these pK_a values:

$CH_3-CH_2-CH_2-COOH$	4.82
$CH_3-CH_2-CHCl-COOH$	2.86
$CH_3-CHCl-CH_2-COOH$	4.05
$CH_2Cl-CH_2-CH_2-COOH$	4.53

These data show that the power of the inductive effect drops off rapidly with distance. 2-Chlorobutanoic acid is much stronger than butanoic acid, but 4-chlorobutanoic acid is hardly any stronger at all:

$$CH_3-CH_2-\underset{\underset{Cl}{\uparrow}}{CH}-COO^- \qquad Cl \leftarrow CH_2-CH_2-CH_2-COO^-$$

effective stabilising influence (Cl is close to $-COO^-$)

almost ineffective (Cl is far away from $-COO^-$)

9.5 Inductive effect of halogen on positive centres

(a) Reactivity of acyl halides

Acyl halides possess a good leaving group in halide ion. The other reason why acyl halides are so readily attacked by nucleophiles is that the inductive effect of the halogen atom *reinforces* the $\delta+$ charge on the carbonyl carbon atom.

e.g.

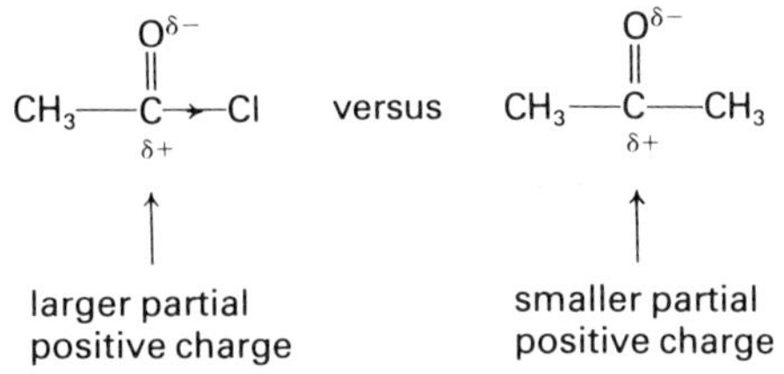

$$CH_3-\overset{\overset{O^{\delta-}}{\|}}{\underset{\delta+}{C}}\rightarrow Cl \quad \text{versus} \quad CH_3-\overset{\overset{O^{\delta-}}{\|}}{\underset{\delta+}{C}}-CH_3$$

larger partial positive charge — smaller partial positive charge

(b) Groups that deactivate benzene towards electrophilic substitution

We saw that alkyl groups with their electron-donating inductive effect *activate* the benzene ring—by the same token, halogen substituents with their electron-withdrawing inductive effect *de*activate the benzene ring.

e.g.

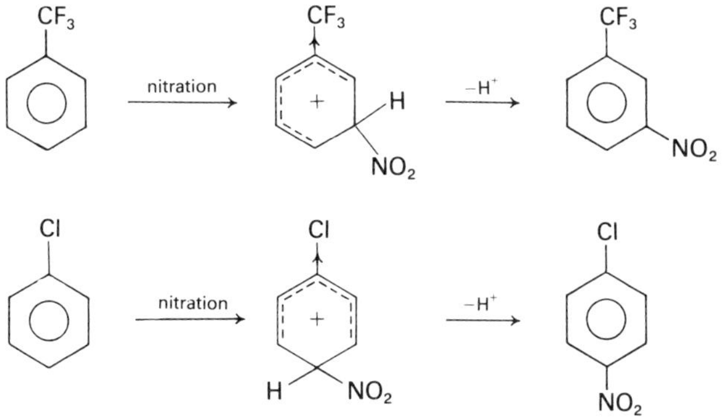

nitration; $-H^+$; CF_3; H; NO_2

nitration; $-H^+$; Cl; H; NO_2

Most electron-withdrawing groups, (e.g. $-CF_3$, $-NO_2$, $-COOH$, $-SO_3H$, etc.), orientate the incoming electrophile to the meta position; halogen substituents themselves ($-Cl$, $-Br$, etc.) are ortho/para directors. We will see the reason for this later [in sections 10.7(d), (e)]. In all cases, however, the inductive effect causes the reaction to be slower or more difficult than with benzene. The intermediate carbocation is destabilised according to category (4) $X \leftarrow Y(+)$ [see section 9.1]. For instance, chlorobenzene is nitrated 33 times more slowly than benzene.

(c) Chloral hydrate

In chloral (trichloroethanal) the two adjacent carbon atoms both carry partial positive charges—because of the polarisation of the carbonyl group and because of the inductive effect of the chlorine atoms:

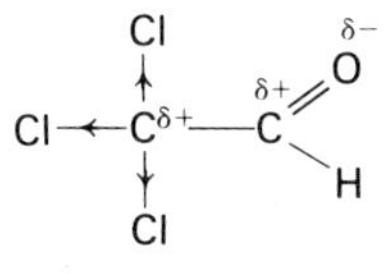

$$Cl_3C^{\delta+}-\overset{\delta+}{C}(=O^{\delta-})H$$

This is clearly an unfavourable situation. The clash of the two δ+ charges can be relieved by the addition of a molecule of water to give the hydrate:

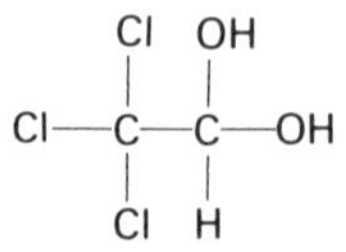

$$\begin{array}{ccccc} & & Cl & & OH \\ & & | & & | \\ Cl & - & C & - & C - OH \\ & & | & & | \\ & & Cl & & H \end{array}$$

In this case therefore (unlike that of most other aldehydes and ketones) the position of equilibrium lies completely over to the side of the hydrate, which is a stable crystalline solid.

(d) Orientation of addition to 3,3,3-trifluoropropene

When hydrogen chloride adds to $CF_3CH{=}CH_2$ the intermediate carbocation would be either $CF_3{-}\overset{+}{C}H{-}CH_3$ or $CF_3{-}CH_2{-}CH_2^+$. You might guess that the first of these would be preferred (it is 2° whilst the other is 1°). However, the first one is in fact *less* stable than the 1° one because of the inductive effect of the $-CF_3$ group:

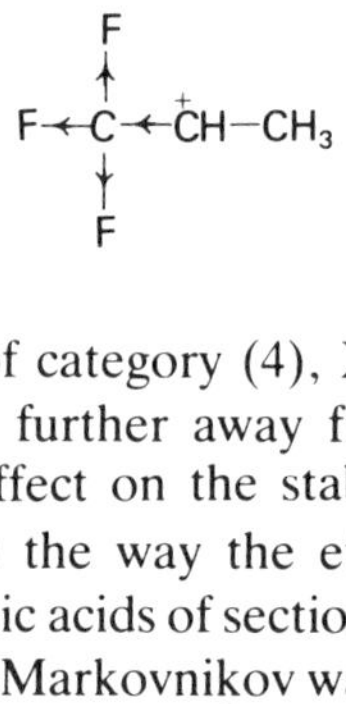

$$\begin{array}{ccccc} & & F & & \\ & & \uparrow & & \\ F & \leftarrow & C & \leftarrow & \overset{+}{C}H-CH_3 \\ & & \downarrow & & \\ & & F & & \end{array}$$

[This is another good example of category (4), X←Y(+). Note that in the 1° carbocation, the $-CF_3$ group is further away from the positive charge and therefore has a much smaller effect on the stability than it does for the 2° carbocation. 'Connect' this with the way the effect of $-Cl$ diminishes with distance in the substituted butanoic acids of section 9.4(c).] Hence, HCl adds to 3,3,3-trifluoropropene in the *anti*-Markovnikov way:

$$CF_3CH{=}CH_2 + HCl \rightarrow CF_3CH_2CH_2Cl$$

9.6 Summary

This chapter exemplifies very well how threads can be found running through organic chemistry that 'connect' seemingly disparate observations. Thus such different things as the rate of nitration of chlorobenzene, the existence of a stable hydrate of chloral, the high acidity of trichloroacetic acid, etc., are all due to the same factor. Similarly, the rate of nitration of toluene, the S_N1 reactivity of *t*-butyl bromide, the relative basicity of ammonia and methanamine, etc., are again all 'connected'.

It is worth realising that many of these factual observations came *before* any theory of explanation. For instance, no one said 'Suppose the methyl group had an inductive effect—let us do some experiments to see if we can substantiate this'. Instead, various experimental observations were made and recorded and published and puzzled over—and eventually someone said 'These things all seem to suggest that a methyl group has an electron-donating tendency'.

10
Resonance

10.1 Introduction

I will assume that you are familiar with the basic concept of resonance, that you know the rules that govern whether a particular resonance structure is important or even possible (some of which we will illustrate with examples as we come to them in this chapter), and that you realise a particular resonance-stabilised species has only *one* actual structure. It does not flip from one resonance form to another, but has a hybrid structure intermediate between the various resonance forms. In connection with the last point, the following analogy may be of assistance to those students who persist in thinking wrongly of a resonance hybrid as a mixture of different species.

Suppose you had a box containing a large number of snooker balls, some blue and some yellow. This illustrates a **mixture**—a mixture of distinctly different objects. Now if you use your imagination and pretend that every so often a blue ball turns into a yellow one, and a yellow one turns into a blue, but the total number of each sort remains constant—then you have an **equilibrium mixture**. If, say, there are many times more blue ones than yellow, we would say the **position of equilibrium** favours blue balls. This would be a pictorial analogy for the mixture of, for example, the keto and enol forms of acetaldehyde (ethanal):

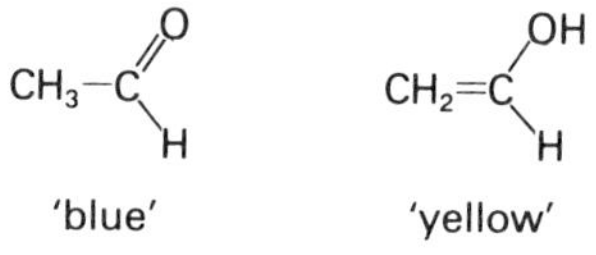

$CH_3-C(=O)H$ 'blue' $\quad$ $CH_2=C(OH)H$ 'yellow'

Now consider a box containing *one sort of object only*—namely pink snooker balls. Pink, of course, is intermediate between red and white, so we might fancifully describe a pink ball as a **resonance hybrid** of a red one and a white one (although as far as this box is concerned, red balls and white balls do not actually exist themselves). A pink ball may be thought of as having some of the character of an imaginary red ball and of an imaginary white ball, but even so it is a real thing in its own right (and it is pink all the time, not red one second and white the next). If our set of pink balls happened to be a delicate shade of very pale pink we might say that *white makes a major contribution to the resonance hybrid*, and red only a minor contribution. Such a picture could be used to illustrate the *single* enolate ion formed by removal of a proton from *either* of the tautomers of acetaldehyde:

$$\underbrace{{}^{-}CH_2{-}C({=}O)H}_{\text{'red'}} \longleftrightarrow \underbrace{CH_2{=}C(O^{-})H}_{\text{'white'}}$$

'pale pink'

The most important thing in connection with the subject of resonance is that **the energy of a molecule or ion that can be represented by more than one structure is lower than the energy that might be estimated for any contributing structure**. (This is a corollary of the point mentioned before in section 1.8 that the more an electrical charge, such as that carried by electrons, can be spread over a wide area the more stable the system is.) In other words, **resonance results in stabilisation**, and the more stable a structure is, the less tendency it has to turn into something else—that is, it is less reactive.

Thus we can summarise in two words what this chapter is going to be all about—the 2 Rs of organic chemistry:

RESONANCE and REACTIVITY

Sometimes, we have a resonance-stabilised **starting material** in a particular reaction, in which case it shows little reactivity—for example, it is difficult to persuade benzene to add chlorine. Sometimes the **product** is resonance-stabilised and then the starting material shows high reactivity—for example, a carboxylic acid readily loses its proton. And sometimes we are concerned with resonance-stabilisation of an **intermediate** in a reaction—for example, phenol is more reactive than benzene to nitration, because in the first case the intermediate (or to be more accurate, the transition state that immediately precedes the intermediate) is more resonance-stabilised than in the second.

In the previous chapter we saw some examples of how inductive effects can affect acidity or basicity. A more common and important factor that can have a bearing on acid or base strength is resonance. Thus we will start by looking at the association between **resonance** and **reactivity** as an acid or a base.

10.2 Resonance and acidity of oxygen acids

(By oxygen acids I mean those compounds in which the acidic proton is lost from an oxygen atom, for example, carboxylic acids, phenols, alcohols, etc.)

(a) Carboxylic acids

We can draw two resonance structures for a carboxylic acid:

$$R{-}C({=}O){-}OH \longleftrightarrow R{-}C(O^-){=}\overset{+}{O}H$$

The first of these is the major contributor; one of the rules of resonance states that a contributor involving **charge separation** is less stable than one that does not. For the carboxylate ion we can also write two structures:

(equivalent to $R{-}C(O^{1/2-})_2$)

This time the two structures are identical and make an equal contribution to the hybrid (which might be drawn as above with half a negative charge on each oxygen). Another important rule of resonance is that **systems described by EQUIVALENT resonance structures have a large resonance-stabilisation**. Thus when a carboxylic acid loses its proton there is a large gain in stabilisation energy. Compare this with the situation of an alcohol losing its proton; neither ROH nor RO^- is resonance-stabilised so we do not have the same driving force for proton loss as in the case of the carboxylic acid. We see therefore how resonance accounts for the much greater acidity of a carboxylic acid ($pK_a \sim 4.8$) than of an alcohol ($pK_a \sim 18$).

(b) Phenols

When phenol acts as an acid it gives rise to the phenoxide ion for which we can write several resonance structures:

You might ask therefore, why phenol is not *more* acidic than a carboxylic acid since there are more contributors to the resonance hybrid? The point is that these resonance structures for the phenoxide ion are *not all equivalent*—the three on the right make only a small (but significant) contribution to the overall structure since in these the negative charge is on carbon rather than on the more electronegative element, oxygen. The net result is that phenol ($pK_a = 9.9$) is intermediate in acidity between an alcohol and a carboxylic acid.

(Unionised phenol itself is also resonance-stabilised to some extent by structures such as:

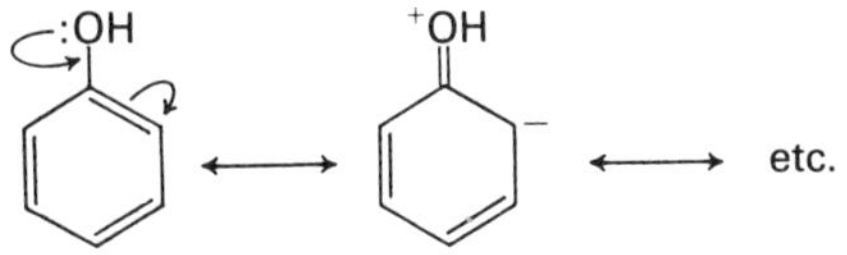

:OH ⟷ $^+$OH ⟷ etc.

Again these forms are of less importance than the equivalent ones for the phenoxide ion because of charge separation. However, these resonance forms indicate that the oxygen in a phenol carries a slight δ+ charge and is therefore less nucleophilic than an alcohol [section 5.5(c)].)

(c) Picric acid (2,4,6-trinitrophenol)

When picric acid ionises, the anion can be described in terms of the same kind of resonance structures as above for phenol, but now in this case there are also important resonance contributors with the negative charge on the various nitro groups:

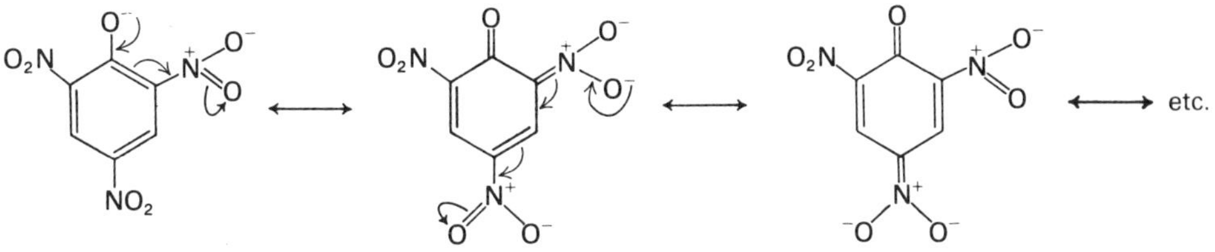

⟷ ⟷ ⟷ etc.

Therefore the anion is much more resonance-stabilised than the simple phenoxide ion and picric acid is thus very willing to give up its proton. The pK_a is 0·4; it is a much stronger acid than, say, acetic (ethanoic) acid. Unlike most phenols, this one would give a vigorous effervescence with sodium bicarbonate solution. (Note that the acidic nature is reflected in the trivial name picric *acid* for what is strictly a phenol.) Analogous compounds with only one or two nitro groups in the ortho and para positions are of correspondingly lesser acidity. Note that a nitro group in the meta position is incapable of stabilising the anion by resonance. We can draw:

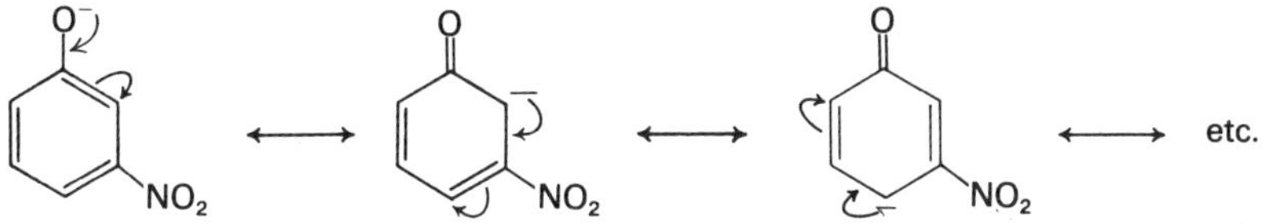

⟷ ⟷ ⟷ etc.

but it is impossible to put the negative charge on the nitro group.

10.3 Resonance and acidity of carbon acids

(a) Triphenylmethane

There is no resonance-stabilisation of course in the case of a simple carbanion such as CH_3^-. However, if the negative charge is adjacent to a benzene ring then the charge can be delocalised and the anion is more stable. The classic example is the triphenylmethide ion:

⟷ etc, etc. (involving the other two benzene rings)

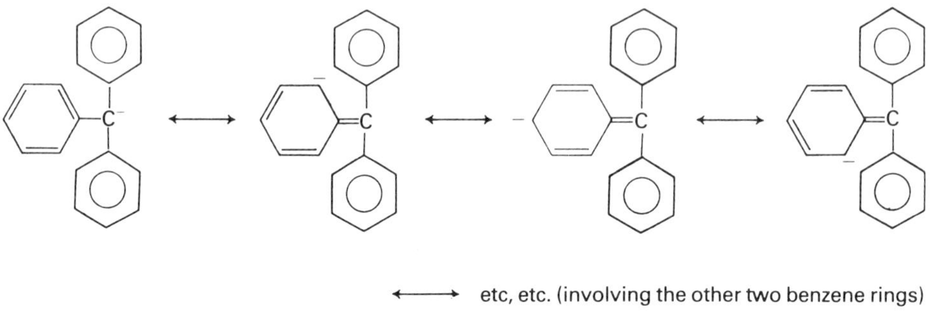

Note that when the negative charge is on one of the benzene rings we lose the usual resonance energy associated with that benzene ring [see section 10.6(a)]. Thus the stabilisation of this anion is not quantitatively great. It accounts for the higher acidity of triphenylmethane (pK_a = 32) than methane (pK_a > 40) but remember that the triphenylmethide ion is still a very strong base.

(b) Cyclopentadiene

1,3-Cyclopentadiene (pK_a = 15) is very much more acidic than triphenylmethane. The reason behind this is the large degree of resonance-stabilisation associated with the *completely symmetrical* delocalisation of charge in the cyclopentadienyl anion:

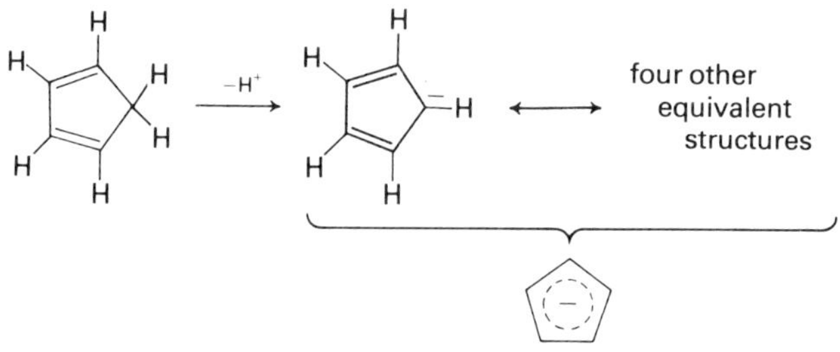

$-H^+$ ⟷ four other equivalent structures

[We will look at this ion again in section 10.6(b).]

(c) Hydrogen alpha to a carbonyl group

As we have seen, the acidity of α-hydrogens is very important; without this there would be no such thing as the aldol or Claisen condensations or any of the many similar reactions. The reason why an α-hydrogen is appreciably acidic in a compound such as propanone (acetone) or ethyl ethanoate (ethyl acetate) is that the resulting anion does not have its negative charge localised on carbon, but spread out over the adjacent oxygen too.

e.g.

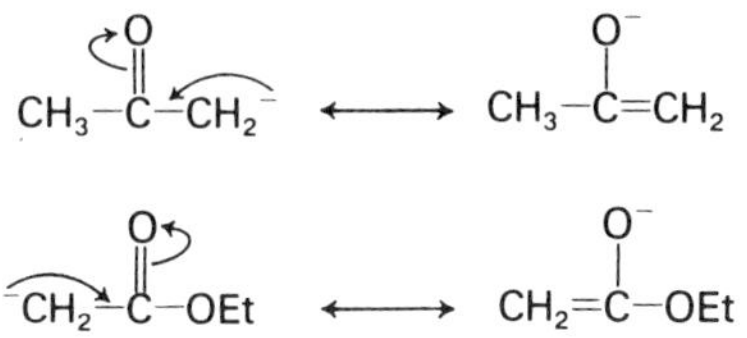

$$CH_3-\overset{O}{\overset{\|}{C}}-CH_2^- \longleftrightarrow CH_3-\overset{O^-}{\overset{|}{C}}=CH_2$$

$$^-CH_2-\overset{O}{\overset{\|}{C}}-OEt \longleftrightarrow CH_2=\overset{O^-}{\overset{|}{C}}-OEt$$

In fact, the *major* contributor to the resonance hybrid in cases such as these is the one with the charge on oxygen; oxygen is more electronegative than carbon and in general is better at accommodating a negative charge. This might prompt you to ask why, when an enolate ion acts as a nucleophile, does it bond itself to the electrophile through its carbon atom rather than oxygen? For instance, in the Claisen condensation why does the first of these processes happen and not the second?

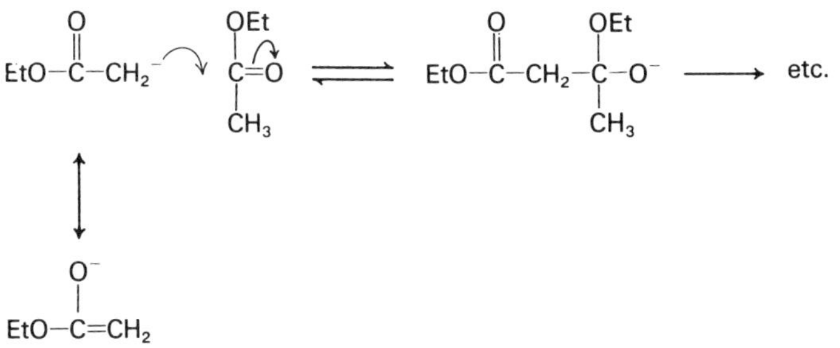

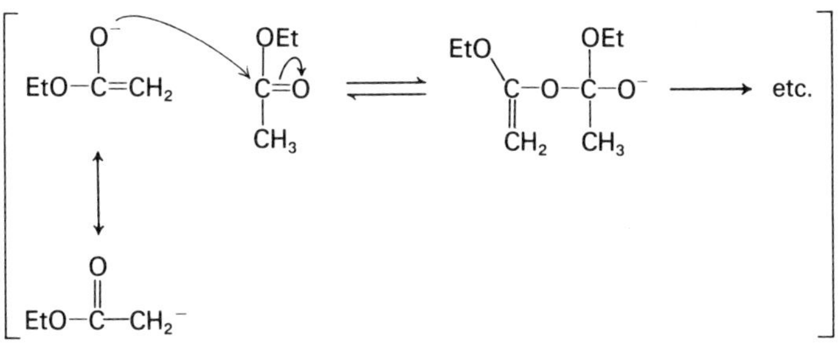

$$EtO-\overset{O}{\overset{\|}{C}}-CH_2^- \; + \; \overset{OEt}{\overset{|}{C}}(=O)-CH_3 \rightleftharpoons EtO-\overset{O}{\overset{\|}{C}}-CH_2-\overset{OEt}{\overset{|}{\underset{CH_3}{\underset{|}{C}}}}-O^- \longrightarrow \text{etc.}$$

$$\updownarrow$$

$$EtO-\overset{O^-}{\overset{|}{C}}=CH_2$$

$$\left[EtO-\overset{O^-}{\overset{|}{C}}=CH_2 \longleftrightarrow EtO-\overset{O}{\overset{\|}{C}}-CH_2^- \right] \; + \; \overset{OEt}{\overset{|}{C}}(=O)-CH_3 \rightleftharpoons \underset{CH_2}{\underset{\|}{C}}(OEt)-O-\overset{OEt}{\overset{|}{\underset{CH_3}{\underset{|}{C}}}}-O^- \longrightarrow \text{etc.}$$

The answer is too complex to go into here. It depends on several factors such as the stability of the products, the rate of the reactions, the solvent, the temperature, etc. Sometimes we *do* get O—C bonds formed, but this is often reversible, and in the majority of cases C—C bond formation dominates. (This is fortunate since as we have seen reactions such as the aldol, Claisen, etc., are important ways of building up the carbon framework in organic syntheses.)

The resonance-stabilisation of the enolate ions accounts for the acidity of propanone ($pK_a = 20$) and ethyl ethanoate ($pK_a = 25$) compared to an alkane ($pK_a > 40$). We can explain the weaker acidity of ethyl ethanoate than propanone as follows. Ethyl ethanoate has a resonance contribution to its structure involving the ethoxy oxygen (which is of course not possible with propanone):

$$CH_3-\overset{\overset{O}{\|}}{C}-OEt \longleftrightarrow CH_3-\overset{\overset{O^-}{|}}{C}=\overset{+}{O}Et$$

This means that the enolate anion of the ester is less stable with respect to the ester (since the ester itself has some resonance-stabilisation) than the enolate anion of the ketone is with respect to the ketone.

The same resonance-stabilisation of an ester compared to a ketone explains why an ester does not react with an organozinc compound in the Reformatsky reaction whereas a ketone does [see section 2.11, 6.5(g)].

(d) Hydrogen alpha to other electron-withdrawing groups

We have seen that hydrogen alpha to several groups other than carbonyl groups is also appreciably acidic [see section 2.5]. *In all cases it is because the resulting anion is resonance-stabilised.*

e.g.

(i)

$$-\overset{\overset{H}{|}}{\underset{|}{C}}-CN \xrightarrow{-H^+} >\overset{-}{C}-C\equiv N \longleftrightarrow >C=C=N^-$$

This means that crossed aldol-type reactions such as the following are possible (think of the details of the mechanism):

$$PhCHO + PhCH_2CN \xrightarrow{NaOEt} Ph-CH=\underset{\underset{Ph}{|}}{C}-CN$$

(ii)

$$-\overset{\overset{H}{|}}{\underset{|}{C}}-NO_2 \xrightarrow{-H^+} >\overset{-}{C}-\overset{+}{N}\begin{smallmatrix}{=O}\\{O^-}\end{smallmatrix} \longleftrightarrow >C=\overset{+}{N}\begin{smallmatrix}{O^-}\\{O^-}\end{smallmatrix}$$

Similarly, this means the following type of reaction is possible (again, can you follow all the steps in the mechanism?):

$$(CH_3)_2CH{-}NO_2 + H_2C{=}O \xrightarrow{OH^-} (CH_3)_2C(NO_2){-}CH_2OH$$

(iii)

$$-\overset{H}{\underset{|}{C}}-\overset{+}{P}(Ph)_3 \xrightarrow{-H^+} \overset{-}{C}-\overset{+}{P}(Ph)_3 \longleftrightarrow C{=}P(Ph)_3$$

The product in this case is an ylide—used in the Wittig reaction [see section 5.6(f)].

(e) Hydrogen alpha to two carbonyl groups

A hydrogen that is alpha to two carbonyl groups is much more acidic than when only one carbonyl group is involved.

e.g.

	pK_a		pK_a
$CH_3COCH_2COCH_3$	9	CH_3COCH_2COOEt	10.7
CH_3COCH_3	20	CH_3COOEt	25

The reason for this of course is that the anion is relatively more stable because there is more scope for resonance.

e.g.

$$CH_3{-}\overset{O}{\overset{\|}{C}}{-}\overset{-}{C}H{-}\overset{O}{\overset{\|}{C}}{-}CH_3 \longleftrightarrow CH_3{-}\overset{O^-}{\overset{|}{C}}{=}CH{-}\overset{O}{\overset{\|}{C}}{-}CH_3 \longleftrightarrow CH_3{-}\overset{O}{\overset{\|}{C}}{-}CH{=}\overset{O^-}{\overset{|}{C}}{-}CH_3$$

(In this case two of the contributors are *identical* which means that there is a large amount of resonance stabilisation energy.)

Recall that the acidity of ethyl 3-oxobutanoate (ethyl acetoacetate) is the factor responsible for pulling the equilibria in the Claisen condensation over to the side of the product [see section 7.6(c)].

It is possible for steric factors to interfere with resonance. For instance, 1,3-cyclohexanedione is just as acidic as pentane-2,4-dione (acetylacetone) and for the same reason:

$$\text{1,3-cyclohexanedione} \xrightarrow{-H^+} \text{anion} \longleftrightarrow \text{enolate} \longleftrightarrow \text{enolate}$$

But the following bridged compound shows little sign of acidity because resonance-stabilisation of the anion is precluded:

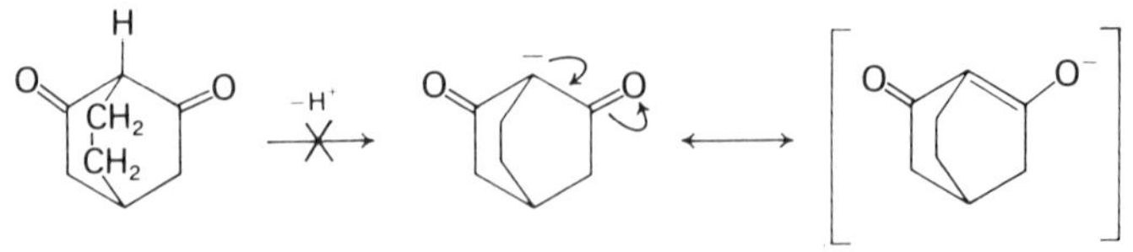

Structures, such as that on the right, with a double bond at the 'bridgehead' position, are too highly strained to contribute at all so there would be no resonance stabilisation of the anion (if it could be formed). Molecular models would be useful here to demonstrate that a planar double bond is geometrically impossible at the bridgehead position.

At this point let us think again about the so-called Doebner reaction of an aldehyde with propanedioic (malonic) acid in the presence of a base such as pyridine [see sections 5.8, 6.5(o)]. This involves as a first step the removal of a proton from malonic acid by the base:

$$HOOC{-}CH_2{-}COOH + B \rightarrow HOOC{-}\overset{-}{C}H{-}COOH + BH^+$$

Why, you may ask, is the proton not lost from one of the acidic carboxyl groups instead? The answer is that to a large extent it *is*. We have an *equilibrium* here between the carboxylate ion and the enolate ion (for which we can draw various resonance structures):

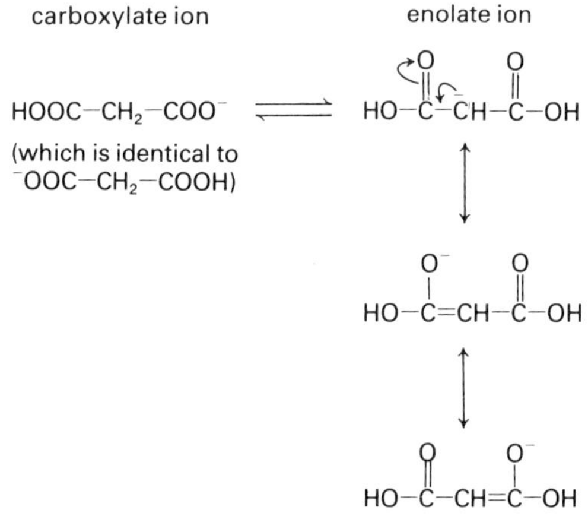

carboxylate ion

$HOOC{-}CH_2{-}COO^-$ (which is identical to $^-OOC{-}CH_2{-}COOH$) $\rightleftharpoons$

enolate ion

$HO{-}C({=}O){-}\overset{-}{C}H{-}C({=}O){-}OH \leftrightarrow HO{-}C(O^-){=}CH{-}C({=}O){-}OH \leftrightarrow HO{-}C({=}O){-}CH{=}C(O^-){-}OH$

This is a good illustration of when to use the symbol $\rightleftharpoons$ (for two or more **different species** that are in equilibrium with each other) and when to use the symbol $\leftrightarrow$ (for various hypothetical resonance structures of a **single** actual species). At equilibrium there is in fact little of the enolate ion present, and moreover in this ion most of the negative charge is delocalised on to the oxygen atoms, and yet the reaction proceeds as if the negative charge were on carbon:

$$(HOOC)_2CH^- + R{-}CH{=}O \longrightarrow (HOOC)_2CH{-}CH(R){-}O^- \longrightarrow \text{etc.}$$

10.4 Resonance and acidity of nitrogen acids

(a) Amides

[We will consider the (lack of) **basicity** of amides in section 10.5(d).] Here we are considering an amide acting as an **acid** (that is, losing a proton):

$$R{-}C({=}O){-}NH_2 \xrightarrow{-H^+} R{-}C({=}O){-}NH^- \longleftrightarrow R{-}C(O^-){=}NH$$

The anion is resonance-stabilised by delocalisation of the negative charge on to the adjacent carbonyl group; thus the proton is lost more readily than from an amine or ammonia (where no resonance is possible). Ammonia is an extremely weak acid (pK_a = 33), but an amide (pK_a = 14–16) is slightly more acidic than, for example, an alcohol ($pK_a \sim 18$).

However, an amide is *not significantly* more acidic than water (pK_a = 15.7) and this means that an amide will *not* dissolve in sodium hydroxide solution by a simple acid/base reaction (as a more acidic compound such as a phenol or a carboxylic acid will). Of course, some simple amides may dissolve in NaOH solution because they are water soluble, and all amides eventually dissolve in boiling NaOH solution as hydrolysis takes place [see section 5.5(t)].

(b) Imides

The acidity of an N—H bond is substantially increased by *two* adjacent carbonyl groups compared to just one (as is the acidity of a C—H bond, of course [see section 10.3(e)]). Phthalimide, for instance, *does* dissolve readily in cold dilute NaOH solution (just like a phenol would):

$$\text{phthalimide (C=O, NH, C=O)} \xrightarrow{NaOH} \text{phthalimide anion } (N^-) \longleftrightarrow (N{=}C{-}O^-) \longleftrightarrow (O^-{-}C{=}N)$$

The resonance structures we can draw for the phthalimide anion are very reminiscent of those we have seen for the enolate ion of pentane-2,4-dione (acetylacetone):

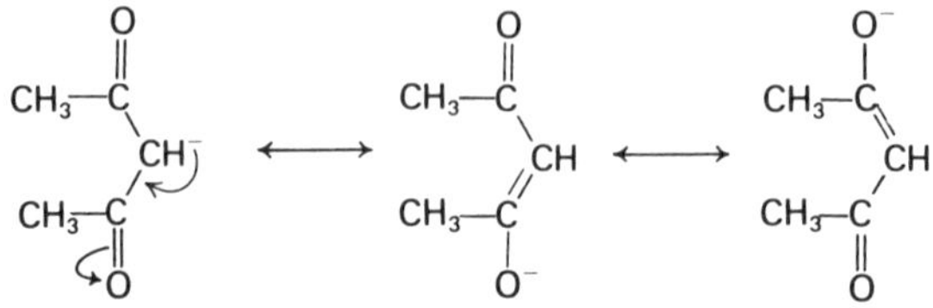

and it is not perhaps surprising that phthalimide and pentane-2,4-dione have the same pK_a (9.0).

(c) Sulphonamides

Sulphonamides are much more acidic than ordinary amides (carboxamides), presumably because there is more scope for resonance delocalisation over *two* S=O bonds rather than just one C=O bond:

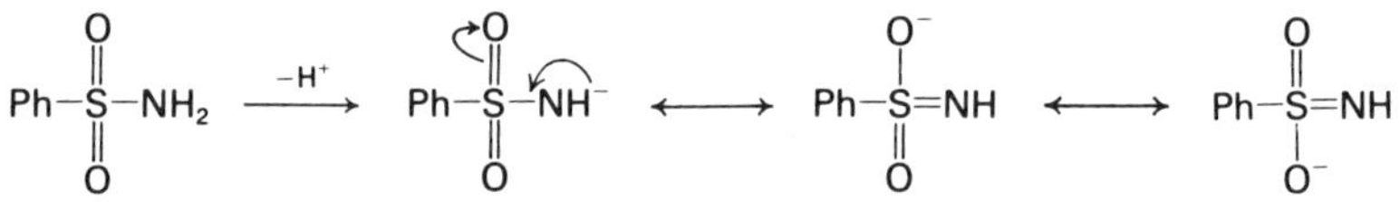

Like phthalimide, benzenesulphonamide (p$K_a \sim 10$) will therefore dissolve in cold dilute NaOH solution, and so will the sulphonamide of a 1° amine. However, the sulphonamide of a 2° amine has no acidic hydrogen and therefore will not dissolve. This is the basis of the Hinsberg method of differentiating classes of amines.

e.g.

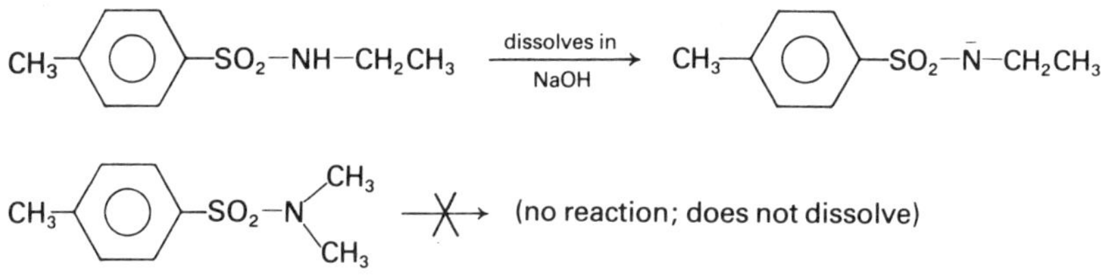

(no reaction; does not dissolve)

10.5 Resonance and basicity

(a) Guanidine

One of the most basic neutral organic compounds (as opposed to an ionic species such as, for example, $RC{\equiv}C^-$) is guanidine, $NH{=}C(NH_2)_2$. As can be seen in

Table 7.1 it is approaching the hydroxide ion in base strength. This high basicity—this great avidity for a proton—is neatly explained by resonance. When guanidine accepts a proton the resulting ion can be represented by three exactly equivalent resonance structures:

$$HN{=}C(NH_2)_2 \xrightarrow{H^+} H_2\overset{+}{N}{=}C(NH_2)_2 \longleftrightarrow H_2N{-}C(NH_2){=}\overset{+}{N}H_2 \longleftrightarrow H_2N{-}C(NH_2){=}\overset{+}{N}H_2$$

We could alternatively represent the ion as:

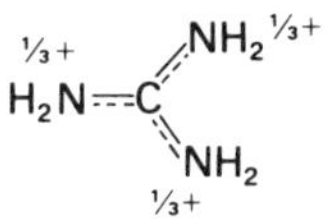

$$\overset{\frac{1}{3}+}{H_2N}{\overset{\cdots}{=}}C(\overset{\frac{1}{3}+}{NH_2})(\overset{\frac{1}{3}+}{NH_2})$$

As we have seen before (for instance with the carboxylate anion [section 10.2(a)]) equivalent resonance structures are associated with a high degree of resonance-stabilisation.

(b) Aromatic amines

After guanidine, the next most basic neutral organic compounds are aliphatic amines. We have seen [section 9.3(d)] that they are of broadly similar basicity to ammonia; there are slight differences due to inductive and solvation effects. Aromatic amines, however, are considerably less basic than ammonia and aliphatic amines, or to put it another way, $Ph{-}NH_3^+$ (pK_a 4.6) is considerably more acidic than $CH_3NH_3^+$ (pK_a 9.64). We can explain this by looking at the various resonance structures we can draw for an aromatic amine such as aniline (benzenamine):

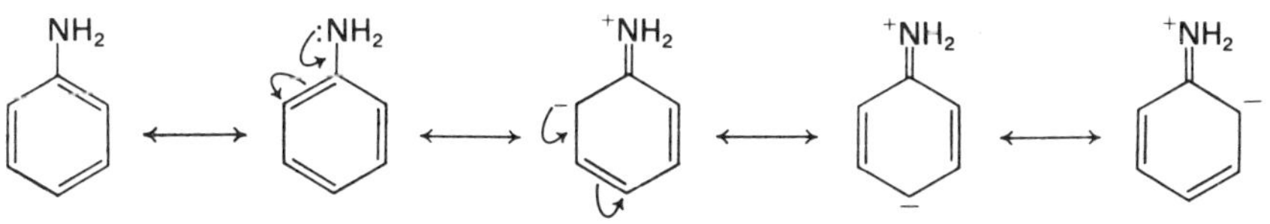

$$C_6H_5NH_2 \longleftrightarrow C_6H_5\ddot{N}H_2 \longleftrightarrow {}^+NH_2{=}C_6H_5^- \longleftrightarrow {}^+NH_2{=}C_6H_5^- \longleftrightarrow {}^+NH_2{=}C_6H_5^-$$

The resonance energy of the benzene ring itself (i.e. that described by resonance between the two Kekulé structures) is 159 kJ mol^{-1}, whereas the overall resonance energy of aniline is 172 kJ mol^{-1}. Thus, the three extra structures on the right above *do not make a quantitatively large contribution* to the resonance hybrid—(they include charge separation and thus are not expected to be very favourable)—but their contribution is *significant*. It means that the nitrogen atom in aniline already carries a small partial positive charge and is thus less ready to accept a proton (that is, less ready to act as a base). Another way of looking at it is to say that the lone pair of electrons on the nitrogen atom is less available to react with a proton because to some extent these electrons are withdrawn into the benzene ring. (Remember that although aniline is less basic than an aliphatic amine it is still basic enough to dissolve in dilute hydrochloric acid.)

One consequence of the weaker basicity of an aromatic amine is that it is not protonated by a carboxylic acid group, but can be protonated by a more strongly acidic sulphonic acid group. Thus, 4-aminobenzoic acid is not zwitterionic, whereas sulphanilic acid (4-aminobenzenesulphonic acid) is. [See also section 7.5.]

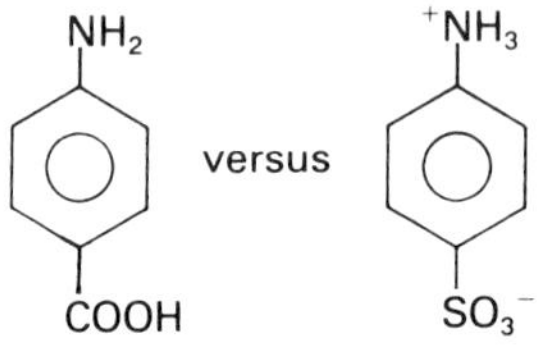

NH_2 / COOH versus $^+NH_3$ / SO_3^-

(c) Picramide (2,4,6-trinitroaniline)

We saw in section 10.2(c) that nitro groups in the ortho and para positions increase the acidity of phenol. By the same mechanism they decrease the basicity of aniline as is clear from considering these resonance structures:

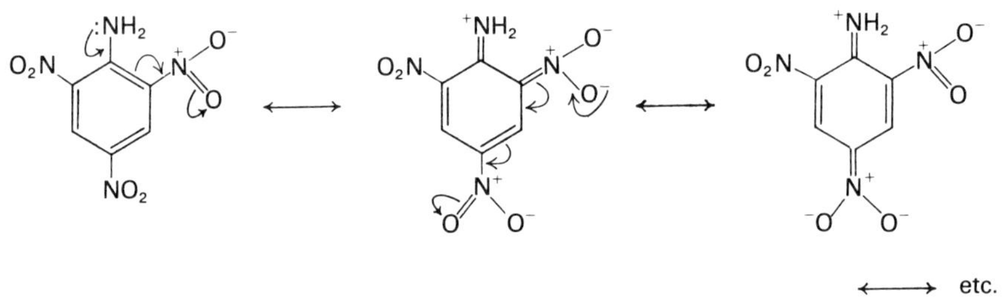

$:NH_2$, O_2N, NO_2 ⟷ $^+NH_2$, O_2N ⟷ $^+NH_2$, O_2N ⟷ etc.

The extra resonance structures involving interaction between the lone pair of electrons on the $-NH_2$ group and the powerfully electron-withdrawing $-NO_2$ groups make a very substantial contribution to the resonance hybrid. In fact, picramide is to all intents and purposes **completely non-basic** (or to put it another way the protonated species, if you could persuade picramide to accept a proton, is an **extremely strong acid** [$pK_a = -9.4$]).

(The lack of basicity is reflected in the trivial name—although structurally this compound is an amine, it is called picramide, since amides are generally less basic than amines [see next section]. In fact this particular amine is much less basic still than the average amide!)

One or two ortho or para nitro groups reduce the basicity of aniline to a proportional extent compared to the three in picramide. However, it is impossible to write resonance structures such as those for picramide when the nitro group is in the meta position. Prove this for yourself.

(d) Amides

Amides can be distinguished from amines since in general amines are soluble in cold dilute hydrochloric acid and amides are not (unless of course they happen to be water soluble). This is explained in terms of the following resonance structures for an amide:

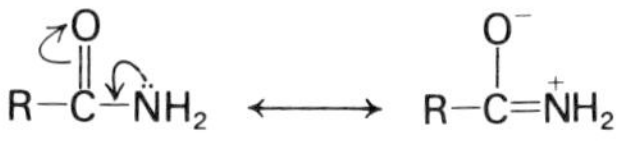

$$R{-}C({=}O){-}\ddot{N}H_2 \longleftrightarrow R{-}C(O^-){=}\overset{+}{N}H_2$$

We can say that the lone pair of electrons on nitrogen is less available to react with a proton, or alternatively we can say that a proton is less likely to react with the N atom since this already carries an appreciable amount of positive charge. In fact, if we force a proton on to the molecule, (e.g. by dissolving it in concentrated sulphuric acid) the site of protonation is probably the oxygen atom:

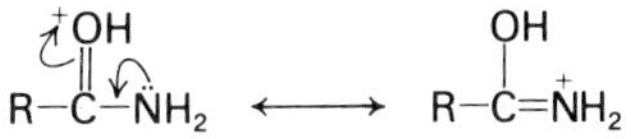

$$R{-}C({=}\overset{+}{O}H){-}\ddot{N}H_2 \longleftrightarrow R{-}C(OH){=}\overset{+}{N}H_2$$

The fact that the resulting cation is resonance stabilised as shown above explains why an amide, although not very basic, is nevertheless more easily protonated than, say, a ketone. [See Table 7.1.]

10.6 Resonance and aromaticity

(a) Benzene

A student is first likely to meet the concept of resonance in the description of the bonding in benzene. Historically, resonance arose as the idea that benzene was a rapidly exchanging **equilibrium** between the two Kekulé structures:

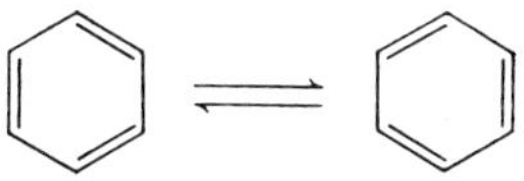

$$\text{Kekulé structure I} \rightleftharpoons \text{Kekulé structure II}$$

(The structures were supposed to be 'resonating' from one form to the other.) Nowadays, of course, we think of the real bonding situation to be intermediate between the two extreme forms (which themselves never exist) and we depict it with the symbol ⟷.

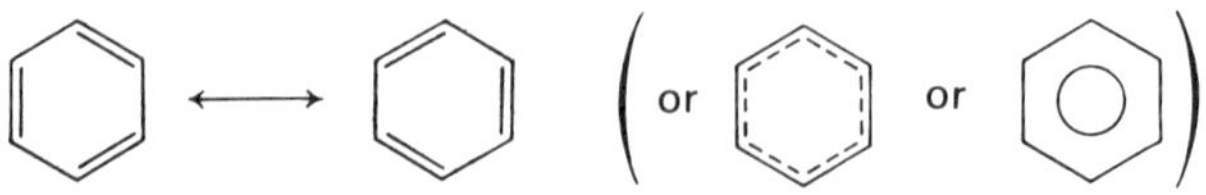

or or

Since the contributors are exactly equivalent there is a large amount of stabilisation energy involved here; this is reflected in the properties of benzene such as resistance to addition, susceptibility to substitution, low heat of hydrogenation, etc., that you should be familiar with (in short, its aromaticity).

The alternative method of describing the aromaticity of benzene is by Molecular Orbital theory which is largely beyond our scope here. Suffice it to say that Molecular Orbital theory is often better at describing electronic structure than resonance is (for example, see the cyclopentadienyl anion in the next section). However, resonance is still very useful in enabling us to rationalise the properties and reactions of benzene and substituted benzenes as will become obvious.

(b) Cyclopentadienyl anion

[See also sections 2.4, 10.3(b).] For this species we can draw five equivalent resonance structures:

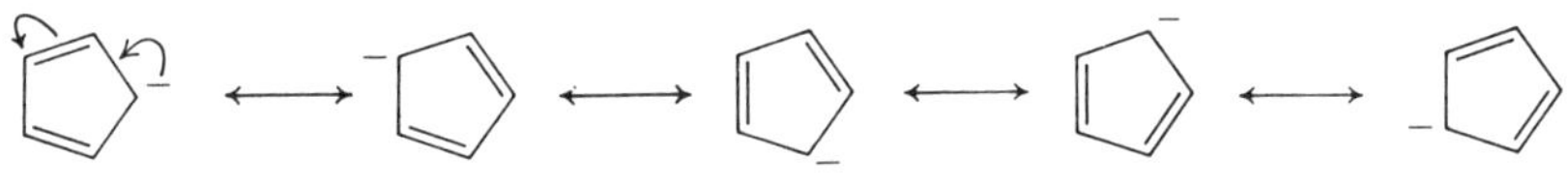

Likewise for the corresponding cation and free radical:

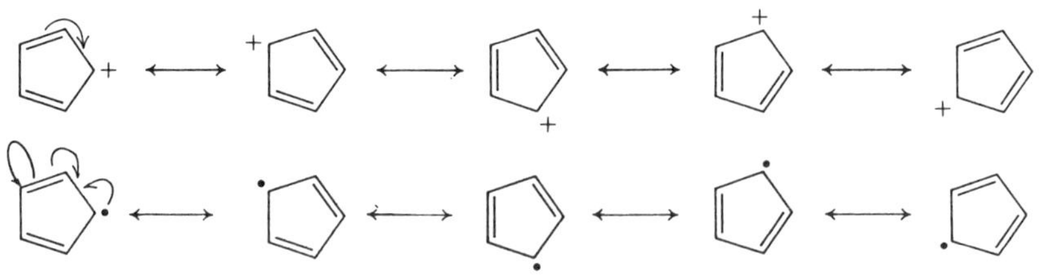

Resonance can tell us nothing more about the relative stabilities of these three species. However, the first one—the anion—is particularly important. In some respects it has an aromatic like degree of stabilisation. The explanation of this is where the Molecular Orbital theory comes into its own. Briefly, the theory predicts that a cyclic unsaturated system with **six** π-electrons [in general, $4n+2$; here $n = 1$] is particularly stable and 'aromatic'. (Benzene of course is the classic example.) The cyclopentadienyl anion has this closed shell of six π-electrons whereas the cation and the free radical have only four and five π-electrons respectively, and are therefore not expected to have such pronounced stability.

(c) Tropylium ion

[See also sections 1.9, 10.10(f).] For the tropylium or cycloheptatrienyl cation we can draw seven equivalent resonance structures:

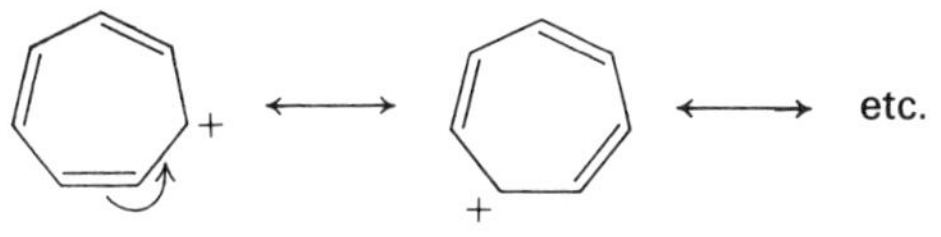

etc.

Indeed, as expected on this basis, the tropylium ion is very stable, but again it is Molecular Orbital theory that explains the stability rather better. Like benzene and the cyclopentadienyl anion, the tropylium ion has the requisite **six** π-electrons for aromaticity to be evident.

10.7 Electrophilic aromatic substitution

(a) General mechanism

After the attack of an electrophile on a benzene ring we can draw three resonance structures for the intermediate positively charged species:

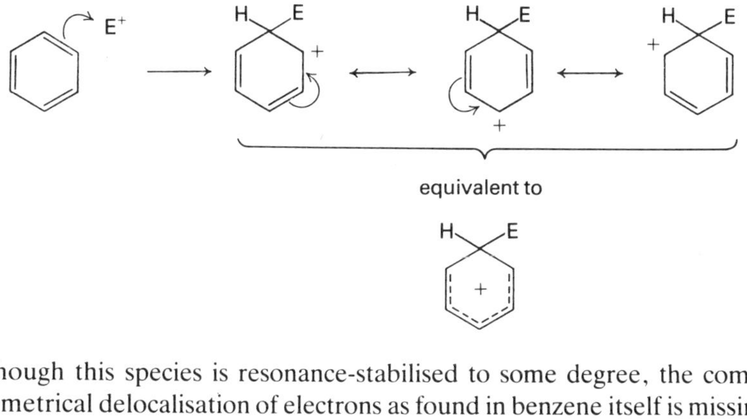

equivalent to

Although this species is resonance-stabilised to some degree, the completely symmetrical delocalisation of electrons as found in benzene itself is missing. It is not surprising, therefore, that the intermediate loses a proton and re-establishes the benzene ring:

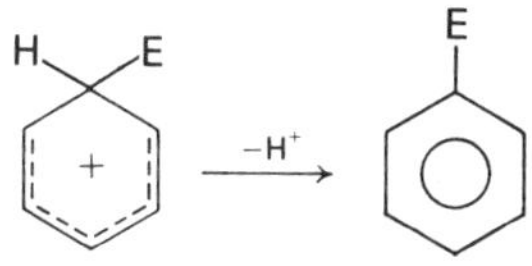

$-H^+$

What we will be interested in next is how the presence of substituents can either stabilise or destabilise the intermediate and thus either accelerate or retard the substitution of electrophiles into the benzene ring. Actually the factor that controls the rate of reaction is the height of the energy barrier en route to formation of the intermediate, but things that stabilise the intermediate will also lower the energy of this barrier, and things that destabilise the intermediate will raise the energy barrier. It is easier to describe the structure of the intermediate than that of the transition state and that is why we concentrate our attention on the intermediate.

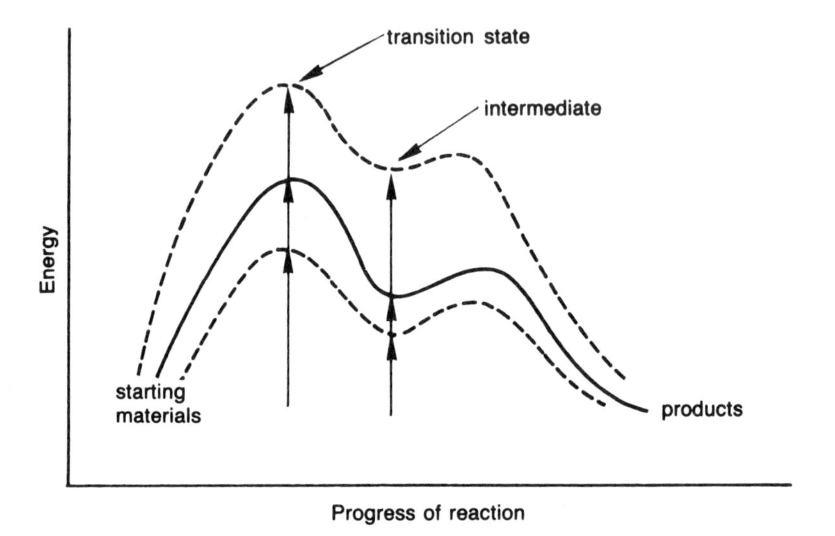

Fig. 8.1

In Fig. 8.1 the solid line shows the progress of an electrophilic substitution reaction for benzene itself. The upper dashed line is for a compound less reactive than benzene, and the lower dashed line is for a compound more reactive than benzene.

(b) Effect of alkyl substituents

Consider the three resonance structures we can draw for the intermediate in the nitration of toluene at the ortho and meta positions:

CH_3, H, NO_2 ⟷ CH_3, H, NO_2 ⟷ CH_3, H, NO_2

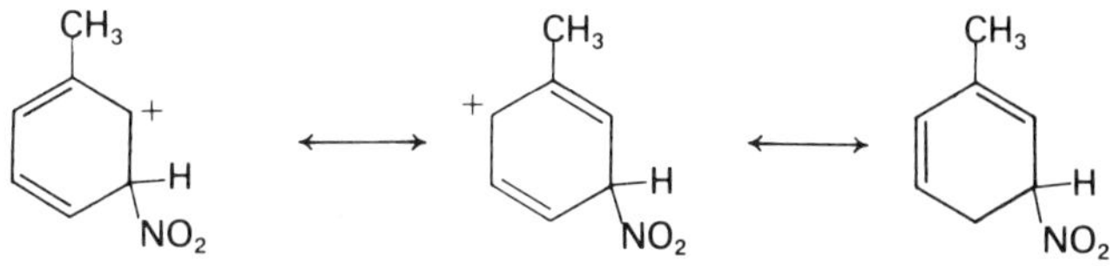

We have seen that the inductive effect of the methyl group activates benzene towards electrophilic substitution [section 9.3(b)]; now we see that for ortho substitution there is a *particularly favourable* resonance contribution in which the inductive effect is brought to bear directly on the positive charge. A similar situation exists for para substitution, which I leave you to draw for yourselves. For meta substitution the positive charge is not shared by the carbon atom carrying the methyl group and so the inductive effect cannot be as effective. Hence alkyl groups are ortho/para directors.

(c) Effect of —OH, —NH_2, —OR, —$NHCOCH_3$, etc.

Consider the resonance structures we can draw for the nitration of phenol at the para and meta positions:

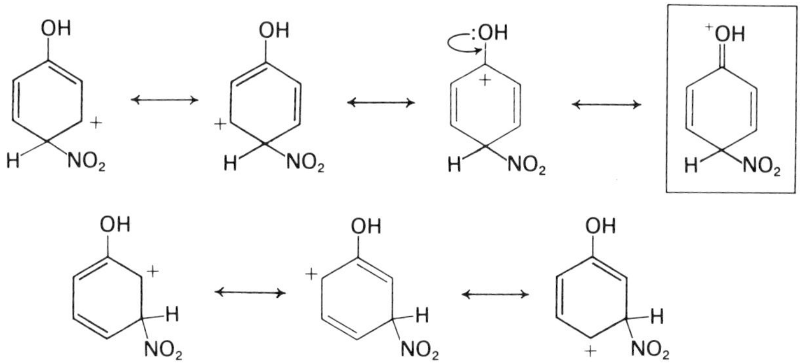

For para attack we have *an extra and particularly favourable contribution*. In this structure all the atoms have full valency shells of electrons, whereas in the others the carbon atom carrying the positive charge has only six electrons in its outer shell. A similar situation arises for ortho attack (prove it for yourself!) but for meta attack this extra resonance form is not possible. Thus, the —OH group is an **activating** group (since it readily supplies electrons to help stabilise the positive charge of the intermediate) and it is an **ortho/para director** (since it specifically stabilises this intermediate for ortho and para substitution).

The same thing applies to many other substituents that, like —OH, have **a lone pair of electrons** on the atom connected to the benzene ring (e.g. $-NH_2$, $-NHR$, $-NR_2$, $-OR$, $-NHCOCH_3$, etc.).

A comparison of the reactivity of a benzene ring carrying the $-NH_2$ and $-NHCOCH_3$ substituents is interesting. The $-NH_2$ group is highly activating—its lone pair of electrons is freely available to stabilise the intermediate positively charged species in the same way as we have seen above for phenol. The result is that aniline reacts with bromine to give 2,4,6-tribromoaniline; the reaction is so facile that it is not possible to limit it to monosubstitution. The reactivity of *N*-phenylethanamide (acetanilide) however is much less, and for the same reason that acetanilide is less basic than aniline [see section 10.5(d)], namely because of resonance in the amide group:

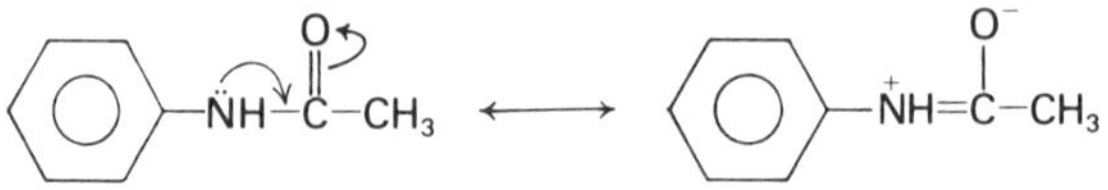

This means that the lone pair of electrons of nitrogen is less available to react with a proton (that is, less basic) and also less available to stabilise the intermediate during electrophilic substitution. The upshot is that the $-NHCOCH_3$ group is still activating (moderately so) and still ortho/para directing (mainly para for steric reasons), but not as activating as the $-NH_2$ group. Hence, monosubstitution of acetanilide is easily achieved.

(d) Effect of halogen substituents

Halogen atoms also have lone pairs of electrons like the substituents in the previous section and therefore we might expect that they too would direct an incoming electrophile to the ortho and para positions. Consider, for example, the following resonance structures and compare them with those in the case of phenol above. (Can you draw the resonance contributors for attack at the meta and para positions?)

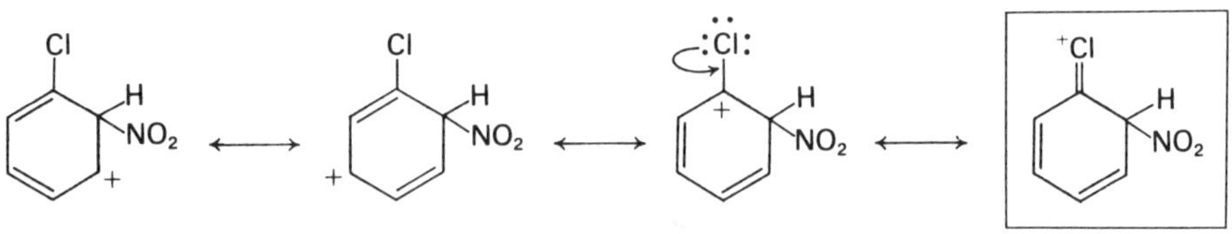

extra resonance form

However, as we have seen [section 9.5(b)], halogen atoms also have a powerful electron-withdrawing inductive effect. On this basis, the third resonance structure drawn above would be a particularly **un**favourable one, since the halogen atom is withdrawing electrons directly from an already positively charged centre:

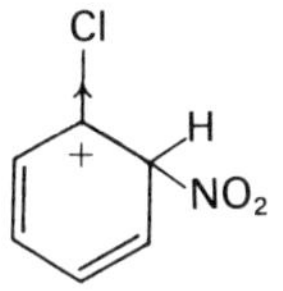

For para substitution there is an analogous unfavourable resonance structure, but not for meta (as you should verify for yourself).

Thus, consideration of chlorine's ability to supply electrons by resonance leads us to expect it to be an activating ortho/para director (like —OH, $-NH_2$, etc.), whereas consideration of chlorine's ability to withdraw electrons by induction tells us it should be a deactivating meta director (like the substituents in the next section). Which is it in fact? Neither—it is a **deactivating ortho/para director**. Thus it appears that it is the **inductive effect** that produces the overall lesser **reactivity** of chlorobenzene (relative to benzene), whereas the **resonance effect** controls the **orientation** of the incoming electrophile. We could perhaps picture the chlorine substituent as trying its hardest to stop the substitution taking place at all, but finally relenting and giving some unwilling assistance at the ortho and para positions.

(e) Effect of $—NO_2$, $—^{+}NR_3$, —COOH, $—SO_3H$, —COR, —CN, etc.

With the exception of halogens, other electron-withdrawing deactivating groups are **meta**-orientating. Consider the resonance structures for the intermediate involved in the nitration of nitrobenzene at the ortho and meta positions: (Para substitution is left for you to examine.)

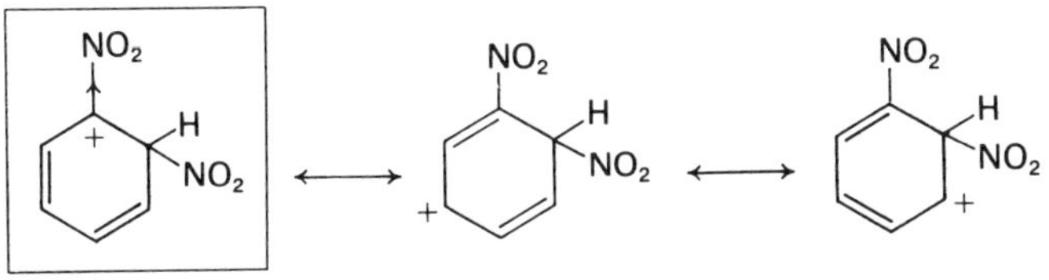

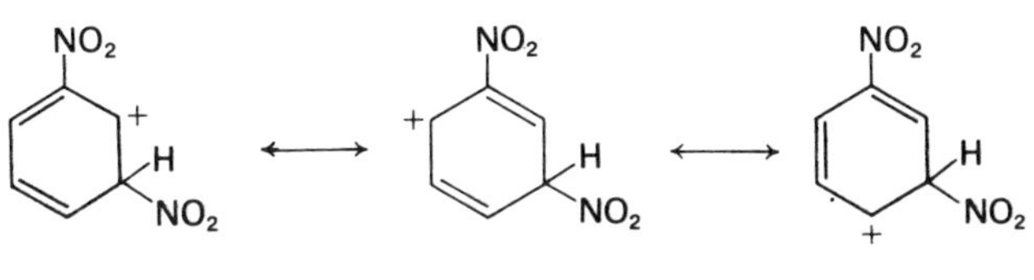

With ortho and para attack, but not meta, there is a *particularly unfavourable* resonance contribution in which the electron-withdrawing inductive effect is brought to bear directly on the carbon carrying the positive charge. (This is the opposite of the case with alkyl groups.) Substitution at even the meta position is made difficult by the nitro group's inductive effect, but not as effectively as at the ortho and para positions. The result is that $-NO_2$ is **de**-activating and **meta**-directing.

The nitro group is electron-withdrawing because the N atom carries a positive charge:

$$R-\overset{+}{N}(=O)-O^- \longleftrightarrow R-\overset{+}{N}(-O^-)=O$$

The other groups in this category also carry a positive or partial positive charge on the atom connected to the benzene ring.

e.g.

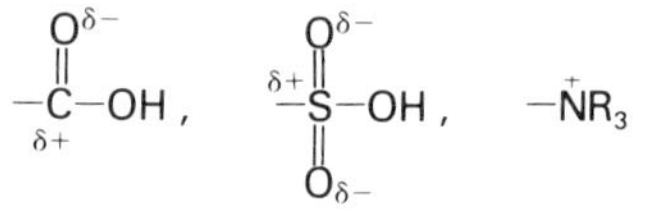

$$-\overset{\delta+}{C}(=O^{\delta-})-OH, \quad -\overset{\delta+}{S}(=O^{\delta-})_2-OH, \quad -\overset{+}{N}R_3$$

(Note that in none of these groups does the atom connected to the benzene ring have a lone pair of electrons. Therefore none of them can promote ortho/para substitution as we saw halogen atoms doing in the previous section.)

(f) Summary

Electrophilic aromatic substitution is a very good illustration of our 'two Rs connection'—resonance and reactivity. The **reactivity** towards electrophilic attack at the ortho, meta and para positions in a whole host of different substituted benzenes can be simply explained in terms of the degree of **resonance-stabilisation** of the intermediate involved.

10.8 Nucleophilic aromatic substitution

(a) Inertness of aryl halides

Alkyl halides readily undergo substitution when attacked by nucleophiles, either by the S_N1 or S_N2 mechanism, depending upon structure and conditions. Not so for simple aryl halides such as chlorobenzene, however, which are inert except under extreme conditions. This can be explained by resonance as follows:

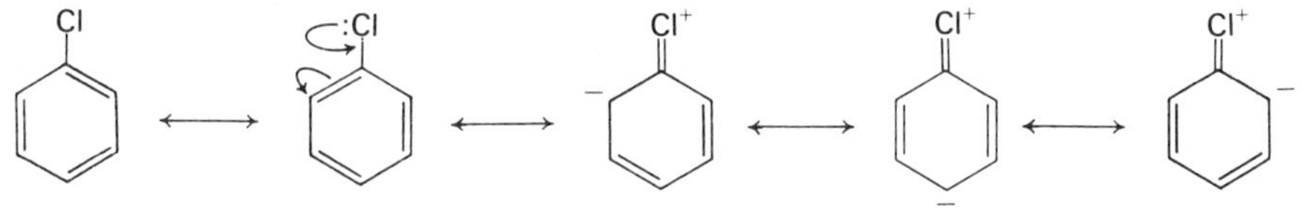

The three structures on the right do not make a quantitatively great contribution to the overall hybrid, but they are sufficiently important to ensure that the carbon to chlorine bond is **partially double** in character and thus **shorter**, **stronger and less reactive** than the C—Cl bond in an alkyl halide.

Connection Study the five resonance structures drawn above and compare them with those for the phenoxide ion [section 10.2(b)] and aniline [section 10.5(b)]. It should be obvious that *exactly the same* kind of resonance consideration explains:

(i) the high acidity of phenol (relative to an alcohol)
(ii) the low basicity of aniline (reactive to an aliphatic amine)
(iii) the low reactivity of chlorobenzene (relative to an alkyl halide).

For similar reasons a vinyl halide is also relatively inert to substitution and elimination reactions:

$$CH_2{=}CH{-}\ddot{C}l \longleftrightarrow {}^{-}CH_2{-}CH{=}\overset{+}{C}l$$

For example, it is more difficult to eliminate HCl from chloroeth**e**ne than from chloroeth**a**ne.

(b) Hydrolysis of picryl chloride (2,4,6-trinitrochlorobenzene)

In marked distinction to what has just been said about the inertness of ordinary aryl halides to nucleophiles, picryl chloride is rapidly hydrolysed in conditions as mild as treatment with warm water:

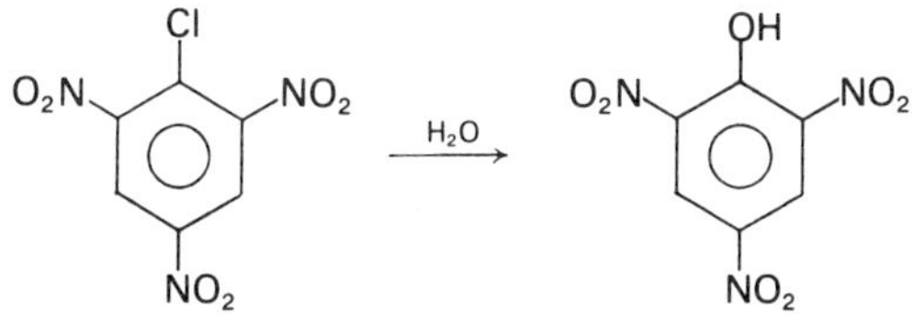

$$\text{2,4,6-}(O_2N)_3C_6H_2Cl \xrightarrow{H_2O} \text{2,4,6-}(O_2N)_3C_6H_2OH$$

The mechanism is neither S_N1 nor S_N2 but is a two step process involving a resonance-stabilised intermediate [see also section 5.9]:

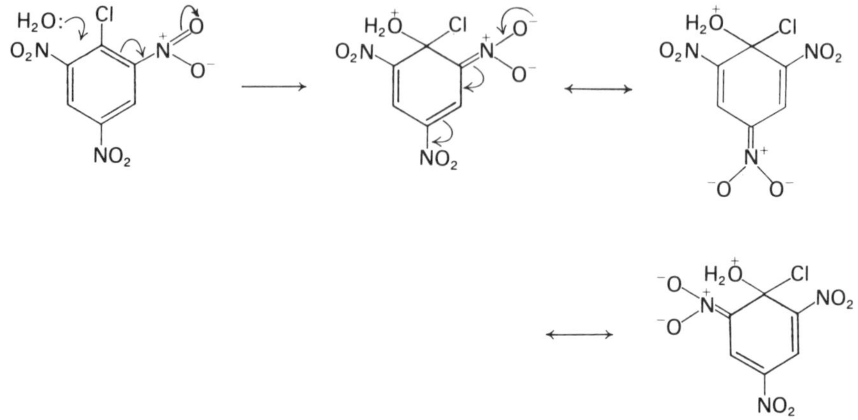

The intermediate then loses Cl^-, and finally H^+, to give the product (picric acid). One or two ortho or para nitro groups activate the chlorine to nucleophilic displacement to a proportionally smaller degree than the three groups in picryl chloride. A meta nitro group is incapable of stabilising the intermediate (as you will see if you attempt to draw the analogous resonance structures).

Note that exactly the same mechanism has been invoked to explain the acidity of picric acid [section 10.2(c)], the lack of basicity of picramide [section 10.5(c)], and now the reactivity of picryl chloride.

> Note also the further '**connection**':
> Electron-withdrawing groups (such as NO_2) **deactivate** the benzene ring towards **electrophilic** substitution, but **activate** the benzene ring towards **nucleophilic** substitution.

10.9 Resonance and diazonium ions

As we saw in our very first example [section 1.2] **aliphatic** diazonium ions are very unstable:

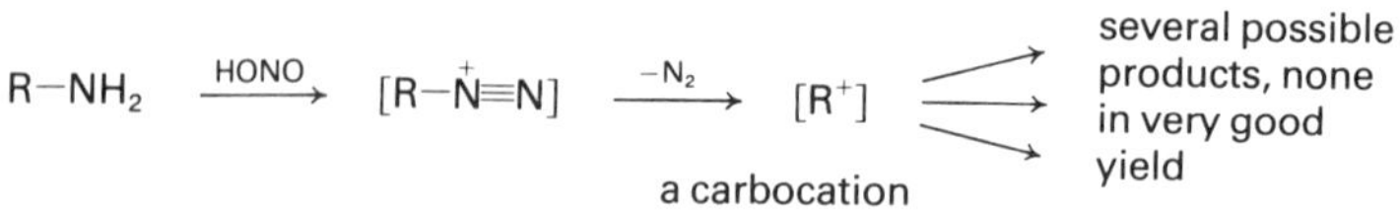

$$R{-}NH_2 \xrightarrow{HONO} [R{-}\overset{+}{N}{\equiv}N] \xrightarrow{-N_2} [R^+] \longrightarrow$$

a carbocation

several possible products, none in very good yield

Aromatic diazonium ions, however, whilst still rather sensitive, are much more stable and more synthetically useful than aliphatic ones. This is because of resonance-stabilisation:

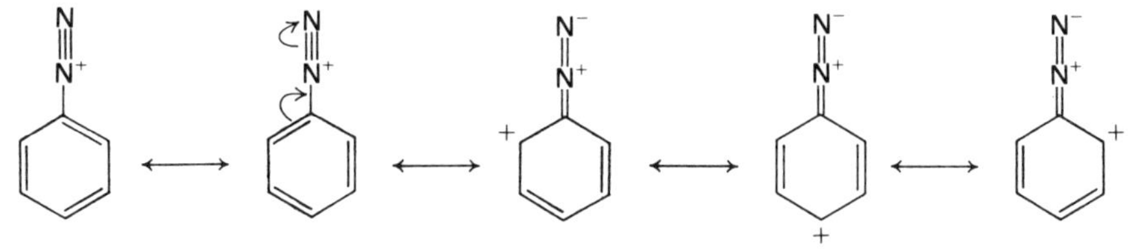

'Connect' this to the similar resonance-stabilisation of the benzyl carbocation [section 10.10(c)].

10.10 Resonance-stabilised carbocations

(a) Hyperconjugation

We have seen that the order of relative stability of alkyl carbocations is 3° > 2° > 1°, and explained it in terms of inductive effects. For example, in the *t*-butyl carbocation there are three methyl groups all tending to 'push' electrons towards the central positive charge [see also sections 1.8, 9.3(a)]:

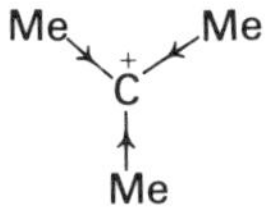

An alternative explanation for the same thing is given by a kind of resonance known as **hyperconjugation**. (Hyperconjugation involves the electrons in sigma bonds; most examples of resonance involve only lone pairs and π-bonds.) On this basis we would write the following structures for the *t*-butyl carbocation:

⟷ Similar structures involving the other two methyl groups.

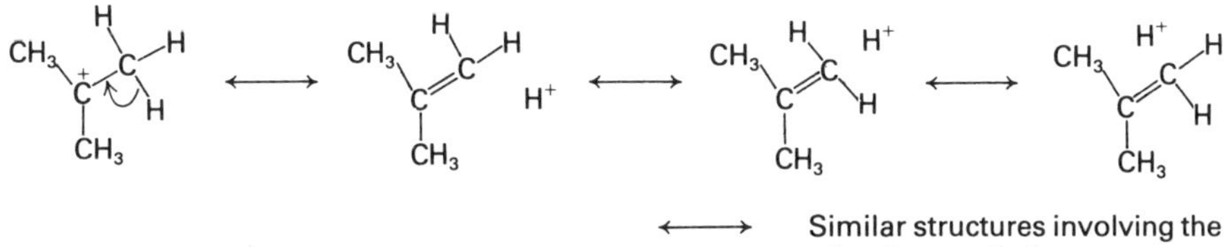

The first one is taken as by far the most important contributor to the hybrid, but the others make a significant contribution (and the more alkyl groups there are, the more scope for resonance, hence explaining the stability order 3° > 2° > 1°). The hyperconjugation structures are not meant to indicate that a proton is ever actually **free**; rather they suggest that taken overall the hydrogens all have a very slight positive charge, the central carbon atom correspondingly has a slightly smaller than integral positive charge, and the carbon–carbon bonds are all slightly double bond in character. Applying this hyperconjugation concept to a cation such as the *t*-butyl one leads us to a very similar mental picture to that gained from looking at it in terms of inductive effects as we did in sections 1.8 and 9.3(a).

(b) Allylic carbocations

A compound such as $CH_3CH{=}CHCH_2Cl$ (1-chlorobut-2-ene) is ostensibly a 1° alkyl halide and yet it is very reactive under conditions that favour the S_N1 mechanism. This is because a resonance-stabilised allylic carbocation is involved:

$$CH_3CH{=}CH{-}CH_2^+ \longleftrightarrow CH_3\overset{+}{C}H{-}CH{=}CH_2$$

(c) Benzylic carbocations

Benzyl halides also show S_N1 reactivity. For the carbocation we can draw these resonance structures:

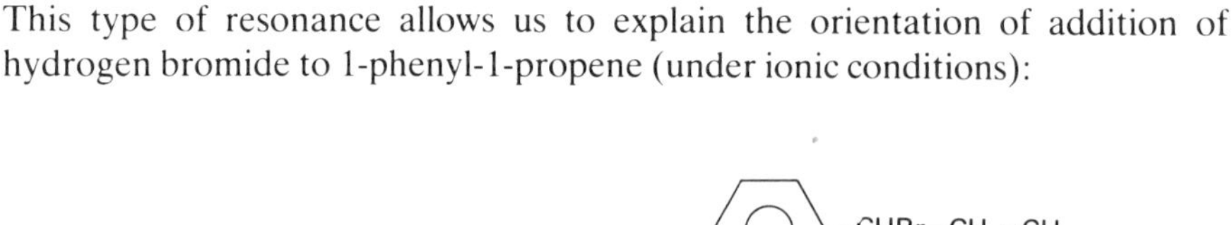

CH_2^+ $\longleftrightarrow$ CH_2^+ $\longleftrightarrow$ CH_2 $\longleftrightarrow$ CH_2 $\longleftrightarrow$ CH_2 + + +

This type of resonance allows us to explain the orientation of addition of hydrogen bromide to 1-phenyl-1-propene (under ionic conditions):

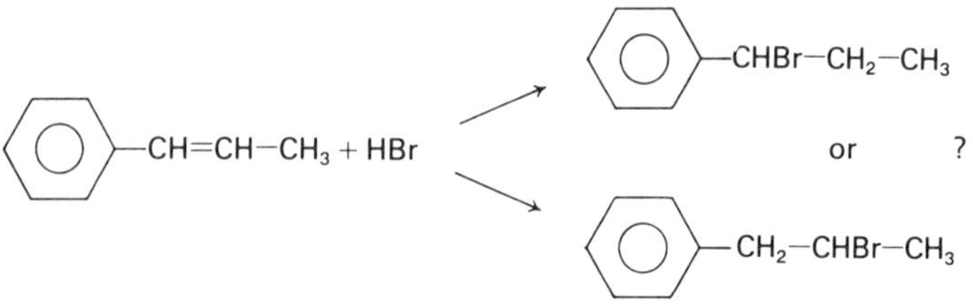

$-CH{=}CH{-}CH_3 + HBr$ → $-CHBr{-}CH_2{-}CH_3$ or $-CH_2{-}CHBr{-}CH_3$?

(Work out which product you would expect before checking back to section 1.11.)

(d) The triphenylmethyl carbocation

With three benzene rings as in Ph_3C^+ the scope for resonance is very much greater than in the benzyl carbocation, $PhCH_2^+$. (The diphenylmethyl carbocation, Ph_2CH^+, is naturally of intermediate stability.) The resonance forms for Ph_3C^+ are:

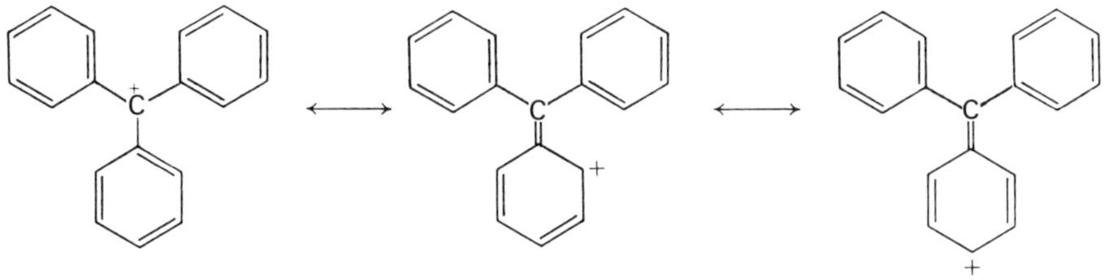

⟷ plus many other similar contributors

The extensive resonance-stabilisation here accounts for the fact that $Ph_3C—Br$ (triphenylmethyl bromide) ionises very readily. Contrast this with the case of the following compound, 1-bromotriptycene, which is extremely resistant to ionisation:

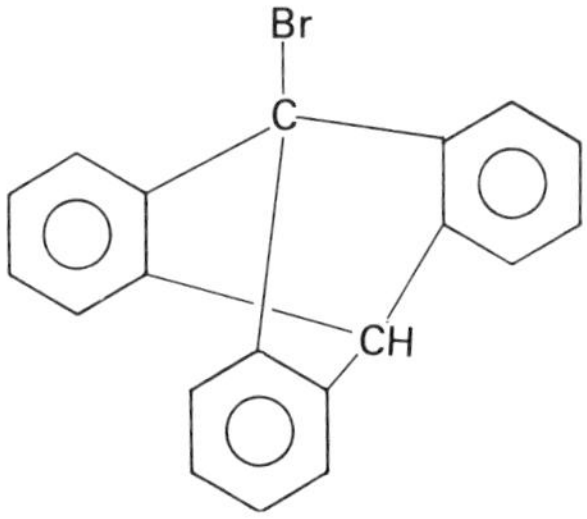

This compound is very similar in structure to $Ph_3C—Br$ (the only difference is that the three benzene rings are tied back by the CH group). Why then should it behave so differently? One reason is that the corresponding carbocation is sterically precluded from being resonance-stabilised in the way that Ph_3C^+ is. For example, the following resonance structure is not permissible:

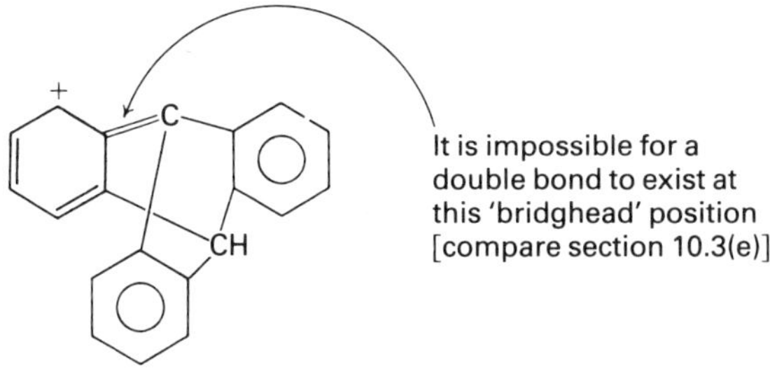

It is impossible for a double bond to exist at this 'bridghead' position [compare section 10.3(e)]

Furthermore, the fact that the carbocation cannot be generated even with the positive charge **localised** at the bridgehead position confirms for us that in general carbocations have to be **planar**:

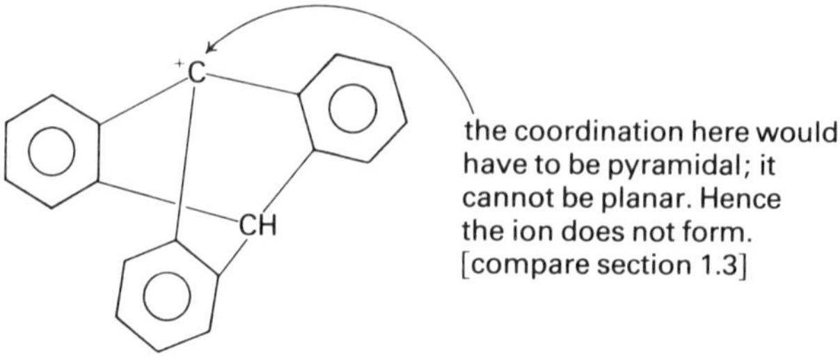

the coordination here would have to be pyramidal; it cannot be planar. Hence the ion does not form. [compare section 1.3]

(Another piece of evidence pointing to the planarity of carbocations is the observation of racemisation in S_N1 reactions as we will see in section 11.3.)

(e) Carbocations-cum-oxonium ions

[See also section 1.9.] We have seen before several cases of resonance-stabilised carbocations in which the predominant contribution has the positive charge not on carbon at all, but on oxygen. For example, a protonated ketone would usually be written:

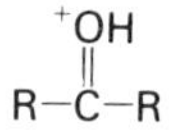

$$R-\overset{\overset{+}{O}H}{\overset{\|}{C}}-R$$

but it has a minor contribution from this carbocation resonance structure:

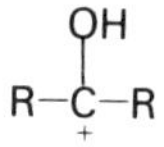

$$R-\overset{OH}{\overset{|}{\underset{+}{C}}}-R$$

Likewise the acylium ion is quite well represented by the structure:

$R-C\equiv O^+$ (perhaps complexed to $AlCl_4^-$)

but in Friedel–Crafts acylation we sometimes see this ion written in its alternative resonance form:

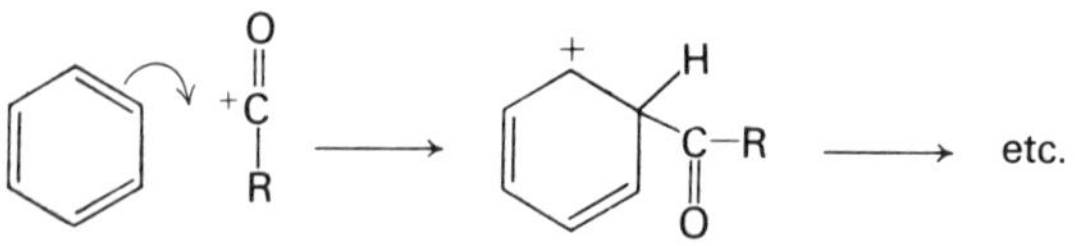

etc.

It is useful to realise that in both resonance structures for a species such as the acylium ion, the oxygen shares an octet of electrons, whilst in one of the resonance forms the carbon only has a sextet:

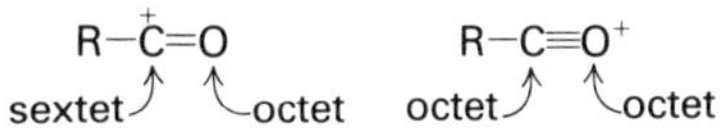

$$\underset{\text{sextet}\quad\text{octet}}{R-\overset{+}{C}=O} \qquad \underset{\text{octet}\quad\text{octet}}{R-C\equiv O^{+}}$$

(The total number of electrons remains constant, of course; only the way they are shared is different.) Thus, it is not surprising that the oxonium resonance structure should predominate.

(f) The tropylium ion

This ion may be described by seven equivalent resonance structures or (rather better) in terms of Molecular Orbital theory [see section 10.6(c)]. Its unusual stability accounts for the fact that tropylium bromide (7-bromo-1,3,5-cycloheptatriene) dissociates more readily than other alkyl halides [see section 1.9]. It is also involved in the following interesting comparison:

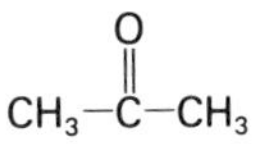

$$CH_3-\overset{O}{\overset{\|}{C}}-CH_3$$

small dipole moment

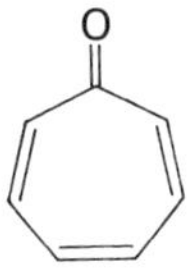

much larger dipole moment

The small dipole moment of the acetone tells us that there is some contribution to the structure from this form:

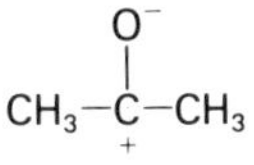

$$CH_3-\underset{+}{\overset{O^-}{\overset{|}{C}}}-CH_3$$

The larger dipole moment of 2,4,6-cycloheptatrienone indicates that the analogous dipolar structure must make a much more significant contribution, the reason being that this form contains within it the very stable tropylium ion:

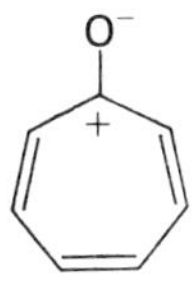

+ 6 other resonance forms

equivalent to

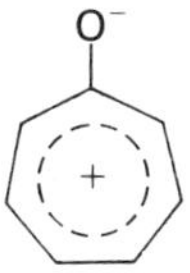

10.11 Resonance-stabilised carbanions

These species have all been dealt with above in looking at the acidity of various organic compounds. *Whenever the hydrogen in a C—H bond is unusually acidic the resulting carbanion is resonance-stabilised* (except for a few stabilised only by inductive effects, such as CCl_3^-). Examples are the triphenylmethide anion, the cyclopentadienyl anion, and numerous varieties of enolate anion.

10.12 Resonance-stabilised free radicals

Primary, secondary and tertiary, allyl and benzyl, triphenylmethyl—the stability of all these free radicals can be explained in *just the same way* as the corresponding carbocations.

(a) Hyperconjugation

We have seen that the order of stability of simple aliphatic free radicals is $3° > 2° > 1°$ (same as for carbocations). This can be elucidated for instance by studying the product ratios of alkane halogenation reactions [section 3.3]. For most practical purposes it is sufficient to accept this order simply as a fact, but sometimes you come across it rationalised in terms of hyperconjugation as follows:

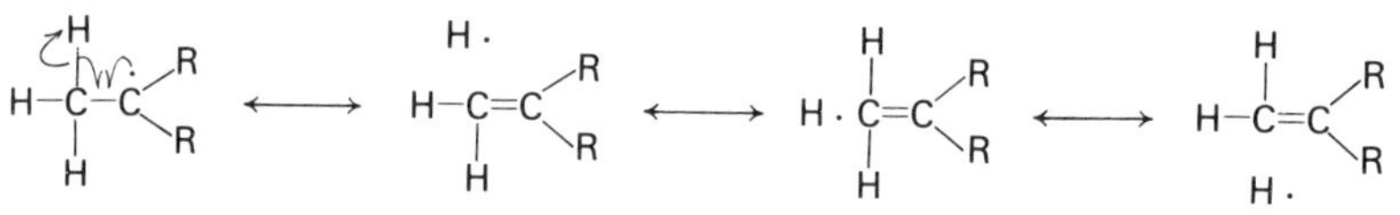

The first structure is the most important contributor—however, the others all help to stabilise the radical by partially delocalising the unpaired electron (and of course the more alkyl groups the more scope for delocalisation).

(b) Allylic and benzylic free radicals

Consider these resonance structures:

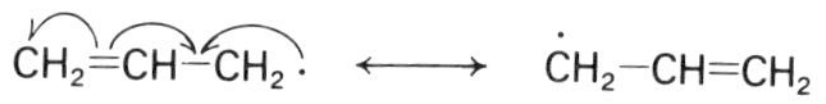

$CH_2{=}CH{-}CH_2\cdot \longleftrightarrow \dot{C}H_2{-}CH{=}CH_2$

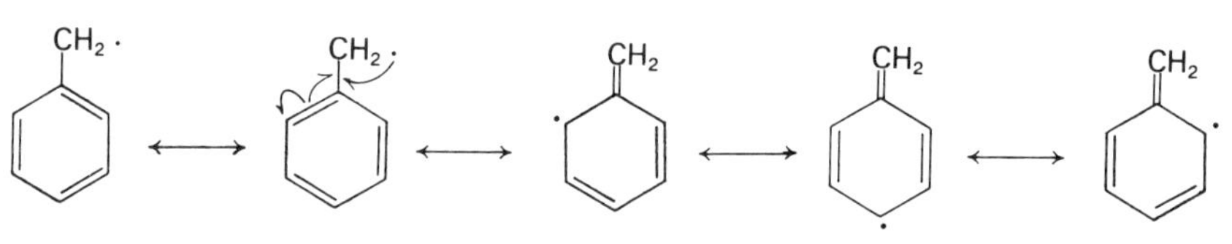

In both cases the unpaired electron is not localised at a single position but spread out over the species as a whole. The consequences of the relative stability of free radicals such as these include the halogenation of propene (and other alkenes) discussed in section 3.5, and the outcome of the addition of hydrogen bromide to 1-phenyl-1-propene (under conditions that promote the free radical mechanism):

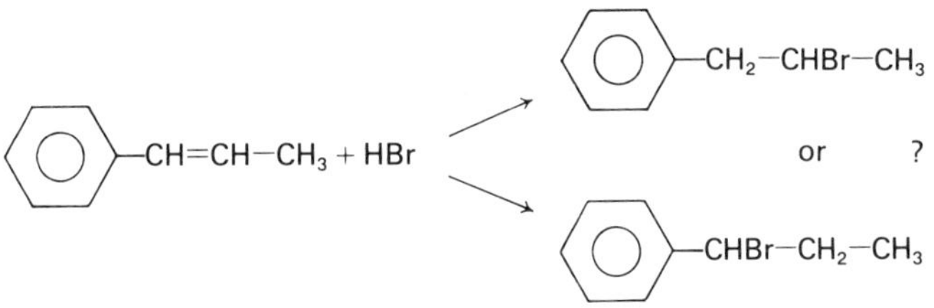

$$C_6H_5-CH{=}CH-CH_3 + HBr \rightarrow C_6H_5-CH_2-CHBr-CH_3 \text{ or } C_6H_5-CHBr-CH_2-CH_3 \ ?$$

(Work out which you would expect before checking back to section 3.7.)

(c) Triphenylmethyl free radical

It was the existence of this particular free radical, $Ph_3C\cdot$ (in equilibrium with its dimer) that was first established (in 1900) and that led to an acceptance that other free radicals were also possible. The unusual stability of this radical is of course associated with very extensive delocalisation of the lone electron. (Try drawing the resonance structures before checking back to section 3.8.)

(d) Diphenylpicrylhydrazyl

One of the most stable free radicals is diphenylpicrylhydrazyl which is deep violet in colour and exists *entirely* as the free radical in the solid state and in solution. Its extraordinary stability is due to extensive delocalisation of the lone electron on to the nitro groups of the picryl residue:

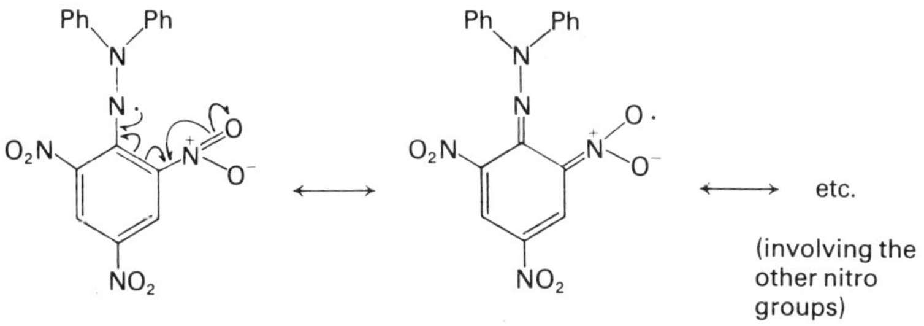

Ph Ph N N· O_2N $\overset{+}{N}$ O O^- NO_2 ⟷ Ph Ph N N O_2N $\overset{+}{N}$ O· O^- NO_2 ⟷ etc. (involving the other nitro groups)

> **Connection** This is the fourth time we have associated the resonance capabilities of the picryl group with various types of reactivity—(i) the acidity of picric acid [section 10.2(c)], (ii) the non-basicity of picramide [section 10.5(c)], (iii) the susceptibility of picryl chloride to hydrolysis [section 10.8(b)], and now (iv) the lack of reactivity of a picryl-containing free radical.

10.13 Résumé

The concept of resonance is a very willing and able work-horse in organic chemistry; without its help the study of properties and reactions of organic compounds would be a very different and less satisfying task. Consider some of the points we have looked at in this chapter:

(i) the greater acidity of a phenol than an alcohol

(ii) the fact that nitrobenzene is substituted at the meta position whereas acetanilide gives ortho and para products

(iii) the high basicity of guanidine

(iv) the solubility of certain sulphonamides in NaOH solution

(v) the acidity of 2,4-pentanedione

(vi) the relative stability of the triphenylmethyl free radical

(vii) the resistance of benzene to addition reactions.

The rationalisation of all these observations and many others involves exactly the same principle—that whenever there is resonance (or in other words whenever electrons or charges are delocalised over a species), then the species is more stable than would be otherwise expected.

11

Stereochemistry

11.1 Introduction

The whole subject of stereochemistry and all its ramifications will not be dealt with here. We will only be concerned with comparing and contrasting a few key types of reaction that have a particular stereochemical outcome. Throughout this chapter the use of molecular models may make the outcome of the reactions easier to follow.

11.2 Free radical halogenation of chiral compounds

Consider the attack of a chlorine atom on one of the stereoisomers of 1-chloro-2-methylpropane:

$$\text{(Me)(Et)(H)C–CH}_2\text{Cl} + \text{Cl}\cdot \longrightarrow ?$$

What happens? From previous considerations [section 3.3] we would expect the chlorine atom to abstract a hydrogen atom preferentially (though not exclusively) from the 3° position, giving HCl and a 3° free radical. The latter species would then react further to give the alkyl halide (**di**halide in this case, 1,2-dichloro-2-methylpropane) and refurnish the chlorine atom:

$$R\cdot + Cl_2 \rightarrow R\text{—}Cl + Cl\cdot$$

But what if you had never been told that this mechanism operates here? You might just as well propose instead that the chlorine atom **ejects**, rather than **abstracts**, the 3° hydrogen atom:

$$R\text{—}H + Cl\cdot \rightarrow R\text{—}Cl + H\cdot$$

And the reaction to refurnish the chlorine atom, so that the chain reaction can continue, would then be:

$$H\cdot + Cl_2 \rightarrow H\text{—}Cl + Cl\cdot$$

An important piece of stereochemical evidence that allows us to choose between these two mechanisms is that the 1,2-dichloro-2-methylpropane isolated from the reaction mixture is found to be **optically inactive**. This would be difficult to explain on the basis of the second mechanism—why should the Cl· eject the H· exactly equally from the front and from the back of the optically active starting material, thus leading to racemisation? (We might expect it more likely to attack from the rear as in an S_N2 process [section 11.4], and thus tend to cause inversion of configuration.) On the other hand, if the first mechanism is followed, and **if the free radical is planar**, then we can easily understand how a racemic product is obtained, following attack by Cl_2 equally well on either side of the free radical intermediate:

Cl—Cl
Me
C—CH₂Cl
Et

Me
C—CH₂Cl
Et
Cl—Cl

(In a similar way, the ionic S_N1 process leads to racemisation of the optically active starting material [section 11.3].) Thus this stereochemical evidence supports the mechanism for free radical halogenation of alkanes that we met in Chapter 3.

11.3 S_N1 reactions

A classic S_N1 mechanism is betrayed by the observation of **racemisation** (if of course the starting material is a single enantiomer and not a racemic mixture to start with, or an achiral compound). The reason for this is that the intermediate carbocation is a planar species [see also sections 1.3, 10.10(d)] and can be attacked by the nucleophile equally well from either side.

e.g.

Me, Et, Pr — C—Br $\xrightarrow{-Br^-}$ Me, Et, Pr — C^+

an *S*-isomer
(optical rotation = *x*°)

$H_2\ddot{O}$ → $H_2\overset{+}{O}$—C(Me)(Et)(Pr) $\xrightarrow{-H^+}$

$\ddot{O}H_2$ → Me, Et, Pr — C—$\overset{+}{O}H_2$ $\xrightarrow{-H^+}$

an equimolar mixture of *S*- and *R*-alcohols
(net optical rotation = 0°)

11.4 S_N2 reactions

In contradistinction to the previous section, a reaction that follows the standard S_N2 mechanism is characterised by **inversion** of configuration.

e.g.

HO⁻ + Me(Et)(H)C–Br ⟶ [HO---C---Br (δ−, δ−; Me, Et, H) (transition state)] ⟶ HO–C(Me)(Et)(H) + Br^-

(*S*-isomer) (transition state) (*R*-isomer)

The nucleophile attacks from the rear of where the leaving group leaves, and the configuration of the chiral centre is inverted (like an umbrella turning inside out in a gale).

Not surprisingly, the observation of **retention** of configuration can often be used to rule out the possibility of an S_N2 reaction. For example, knowing that optically active 2-bromobutane undergoes inversion of configuration when reacted with hydroxide ion (see above) might lead you to think that 2-butyl benzoate would react similarly (with the benzoate ion acting as leaving group instead of Br^-):

HO⁻ + Me(Et)(H)C–O–C(=O)–Ph ⟶ HO–C(Me)(Et)(H) + ⁻OOC–Ph

However, if 2-butyl benzoate is made from benzoyl chloride and *S*-2-butanol and then hydrolysed, it is *S*-2-butanol that is recovered, not *R*-2-butanol. The S_N2 mechanism above is therefore excluded, and we propose instead the mechanism already covered in section 5.5(n), involving this process:

Me(Et)(H)C–O–C(=O)–Ph + HO⁻ ⟶ Me(Et)(H)C–O–C(O⁻)(OH)–Ph ⟶ Me(Et)(H)C–O⁻ + HO–C(=O)–Ph ⟶ Me(Et)(H)C–OH + PhCOO⁻

Retention of configuration does not always exclude the S_N2 mechanism; it may disguise the sequential occurrence of two S_N2-type processes. For example, optically active 2-bromopropanoic acid is hydrolysed in dilute NaOH with retention of configuration and this is thought to be because of the operation of this mechanism, involving the intermediacy of a lactone (or cyclic ester):

$$CH_3-\underset{Br}{\overset{H}{C}}-COOH \xrightarrow{HO^-} CH_3-\underset{Br}{\overset{H}{C}}-C(=O)O^- \xrightarrow{A} \text{lactone} \xrightarrow{B} CH_3-\underset{OH}{\overset{H}{C}}-COO^-$$

Reaction A, an S_Ni (or intramolecular nucleophilic substitution reaction) goes with inversion of configuration, as does reaction B (an S_N2); inversion and re-inversion result in retention. (In more concentrated hydroxide solution, Br^- tends to be displaced directly by attack from OH^- in a normal S_N2 reaction with the expected inversion of configuration.)

11.5 Reactions intermediate between S_N1 and S_N2

We do not always observe clear-cut racemisation or inversion of optically active reactants. Sometimes, both processes happen to an intermediate degree. For instance, some 2° alkyl halides might be thought of as simultaneously reacting in both the S_N1 and S_N2 modes (whereas we know that 1° alkyl halides generally favour the S_N2 and 3° ones the S_N1 mechanism). An alternative, and probably more realistic way of looking at cases like this is to consider the formation of **ion pairs**. Suppose an alkyl halide (or similar compound) ionises but does not fully dissociate:

$$XYZC-Br \rightleftharpoons XYZC^+ \cdots Br^-$$

an ion-pair; the leaving group has not yet fully left.

In this case attack of the nucleophile is *not* equally easy from either side of the carbocation:

$$Nuc^- \rightarrow XYZC^+ \cdots Br^- \leftarrow Nuc^-$$

Attack unhindered; predominates but not exclusively

Attack hindered but not completely ruled out.

The net result as far as stereochemistry is concerned is clearly somewhere *between* total inversion and racemisation. Thus we see that the classic S_N1 and S_N2 processes are the extreme ends of a spectrum of possible reaction mechanism, and quite often practical examples of nucleophilic substitution reactions do not fall cleanly into either of these two categories.

An instructive example is found in halohydrin formation from alkenes and halogens in the presence of water.

e.g.

$$CH_3CH{=}CH_2 \xrightarrow{Cl_2/H_2O} CH_3{-}CHOH{-}CH_2Cl$$

$$(\text{not } CH_3{-}CHCl{-}CH_2OH)$$

We might think that the first step would be attack of chlorine on the alkene to give a carbocation and Cl^-, and in this case there would be two possibilities for the carbocation:

$$CH_3CH{=}CH_2 + Cl_2 \begin{cases} \xrightarrow{\text{either}} CH_3{-}\overset{+}{C}H{-}CH_2Cl + Cl^- \\ \xrightarrow{\text{or}} CH_3{-}CHCl{-}CH_2^+ + Cl^- \end{cases}$$

The second step would be attack of water on the carbocation in an S_N1-like process [see section 11.3], and we would expect reaction to favour the more stable 2° rather than the 1° carbocation:

$$\underset{H_2O:\curvearrowright}{CH_3{-}\overset{+}{C}H{-}CH_2Cl} \longrightarrow CH_3{-}\underset{H_2O^+}{\underset{|}{C}H}{-}CH_2Cl \xrightarrow{-H^+} CH_3{-}\underset{OH}{\underset{|}{C}H}{-}CH_2Cl$$

This mechanism, therefore, would nicely account for the observed product. However, as will shortly be explained [section 11.8(a)], it does *not* account for the stereochemistry of the halogenation of an alkene such as *cis*-2-butene. To do this it is necessary to propose the intermediacy of a **cyclic halonium ion**. Returning to chlorohydrin formation from propene, therefore, we would have nucleophilic attack by water at one end or other of the following species:

$$CH_3{-}CH\overset{\overset{+}{Cl}}{\diagup\!\!\diagdown}CH_2 \quad (a) \leftarrow H_2\ddot{O} \rightarrow (b)$$

For this essentially S_N2 process we would expect attack to predominate at the less hindered 1° position as in (b) above, but it is attack at position (a) that actually leads on to product. How do we solve this problem? By considering the cyclic chloronium ion as a resonance hybrid of the following contributors:

$$CH_3{-}CH\overset{Cl^+}{\diagup\!\!\diagdown}CH_2 \longleftrightarrow CH_3{-}\overset{+}{C}H\overset{Cl}{\,\,\diagdown}CH_2 \longleftrightarrow CH_3{-}CH\overset{Cl}{\diagup\,\,}CH_2^+$$

The second of these (as a 2° carbocation) makes a greater contribution to the hybrid than the last one (as a 1° carbocation), and so we can understand why water prefers to attack at position (a) above even though it is more sterically hindered. In other words the reaction mechanism could be described as **S_N2 with considerable S_N1 character**.

11.6 E2 reactions

Usually, the stereochemistry of the E2 elimination mechanism is that described by the phrase **'anti periplanar'**. Periplanar signifies that, in the transition state, the hydrogen that is abstracted by base, the two carbon atoms that end up linked by a double bond, and the leaving group are **all in the same plane**. **Anti** implies that the hydrogen and the leaving group are on **opposite** sides of the two carbon atoms. This is illustrated by:

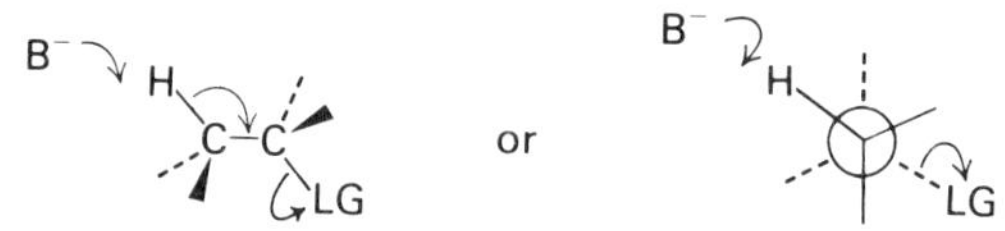

B⁻ H C–C LG or B⁻ H LG

B = Base
LG = Leaving Group

The Newman projection formula (on the right) accentuates that with an *anti* periplanar arrangement of the groups involved all the substituents are **staggered** (which is a lower energy situation than if they were **eclipsed**) and this is presumably one of the main reasons why the reaction mechanism has this stereochemistry. Furthermore, in the course of the reaction the electron movement towards the α-carbon is effectively from the rear of the position from which the leaving group departs—just as in the S_N2 mechanism.

Let us look at an example where this reaction mechanism leads stereospecifically to a particular product. Consider some of the various conformations of a particular stereoisomer of 1-bromo-1,2-diphenylpropane:

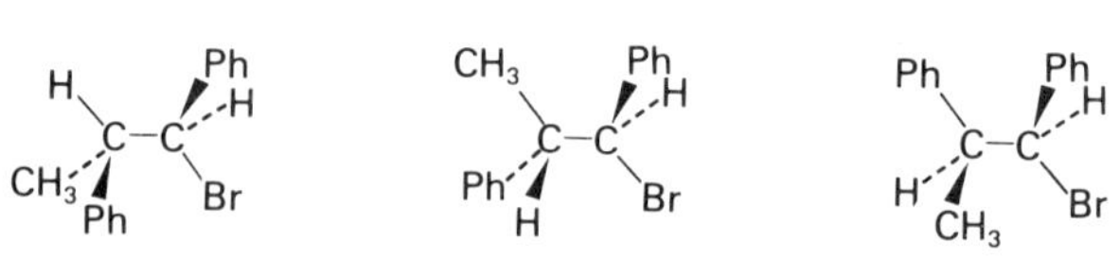

H, CH_3, Ph – C–C – Ph, H, Br CH_3, Ph, H – C–C – Ph, H, Br   Ph, H, CH_3 – C–C – Ph, H, Br

It must be stressed that these are just different conformations of the *same* structure; the central C–C bond is free to rotate and the molecules will distribute themselves over these various conformations depending on their relative stability. Just from a quick look at these drawings it is impossible to predict whether the resultant alkene (when HBr is eliminated) will have the two phenyl groups on the same side or on opposite sides of the double bond.

However, only one of these conformations has the H and Br (that are lost) in the *anti* periplanar arrangement. Thus we can follow the mechanism through this particular conformation to see that the alkene product is in fact the *cis* isomer in this case:

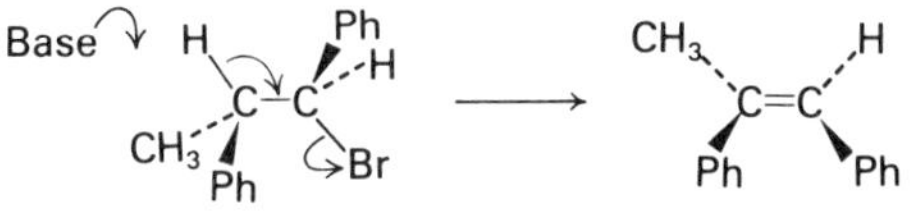

Now consider elimination of HCl from 2-chlorobutane. Unlike the preceding case there are *two* conformations that have an *anti* periplanar arrangement of the substituents that are lost; they are (in Newman projection, with the bulkier groups circled):

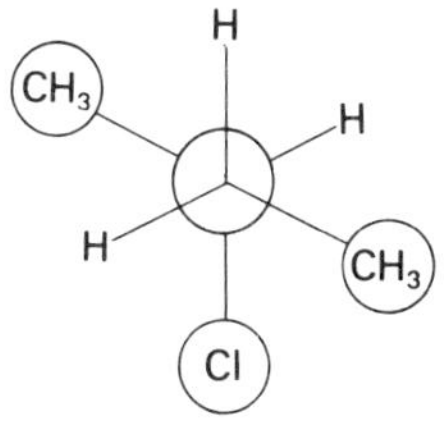

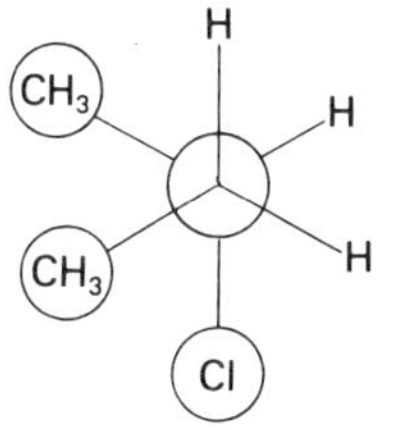

The one on the left is of lower energy because it has less steric interactions between the three bulkier substituents. Loss of HCl from this conformation would clearly result in *trans*-2-butene, rather than the *cis* isomer, and this is indeed what is observed in practice to be the predominant product (by a ratio of 6 to 1). However, there is only a relatively low energy barrier to interconversion of the two conformers drawn above and the slight preference for one of them does not really explain why we get such a preponderance of *trans* over *cis* product. What we should properly consider here are the **transition states** rather than the starting material. The two possibilities are drawn below with the H partially removed by the base, the Cl partially departed, and the double bond partially formed (but not visible in Newman formulae):

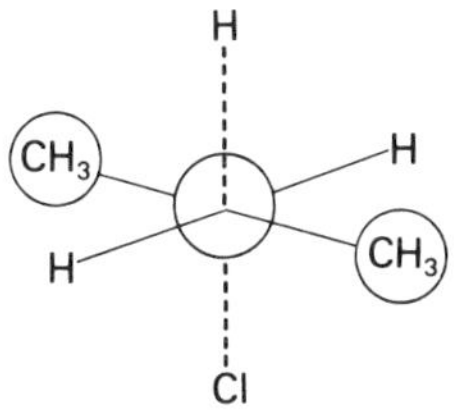

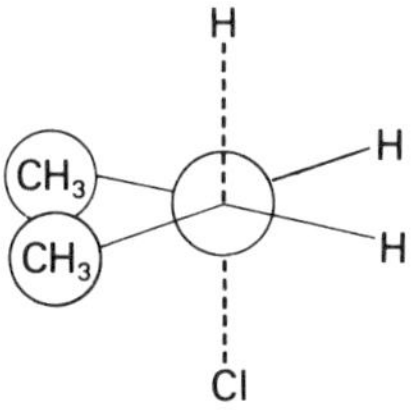

In these transition states the four substituents that remain behind on the alkene are part way to taking up their final planar arrangement. In the one on the right, the two methyls are almost eclipsed; thus on this basis we can safely conclude that the transition state on the left will be preferred, explaining why *trans*-2-butene is the major product. The energy difference between the two transition states is presumably slightly less than the energy difference between the two completely planar alkenes. Measurement of heats of hydrogenation shows that the *trans* alkene is more stable than the *cis* by 4.2 kJ mol^{-1}.

[Although we will not go into this any further, sometimes (especially with certain cyclic compounds) the stereochemistry of the E2 mechanism is *syn* periplanar rather than *anti*. Here in the transition state, the hydrogen and the leaving group are on the same side of the C—C bond (and eclipsed).]

11.7 *Syn* additions to alkenes and alkynes

(a) Addition of $KMnO_4$ and OsO_4 to alkenes

These reagents form cyclic derivatives with alkenes that are then hydrolysed to vicinal diols (or glycols):

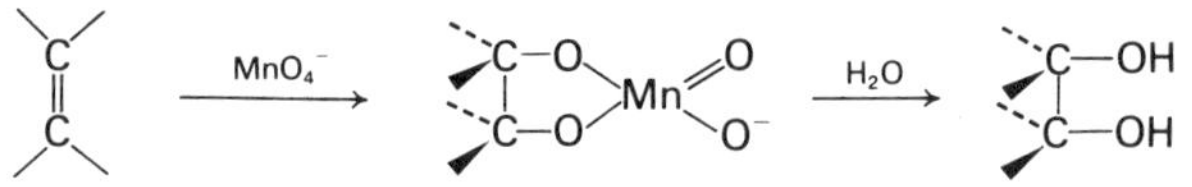

MnO_4^- → H_2O →

Both OH groups are obviously added effectively from the *same* side of the original double bond, (i.e. a *syn* addition). Therefore, *cis*-2-butene, for example, leads to the *meso* glycol:

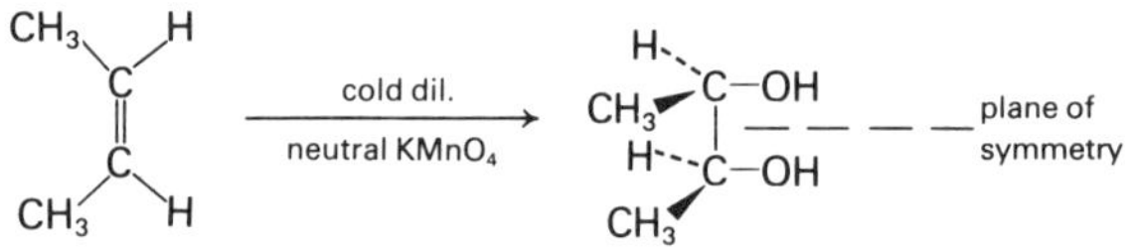

cold dil. neutral $KMnO_4$ → plane of symmetry

Do not forget that the C—C single bond in this product is free to rotate and the molecule has an infinite variety of conformations. For example, the following is still the same meso diol:

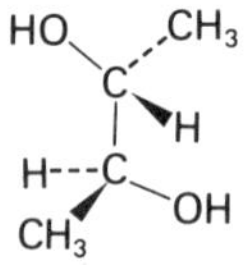

(b) Addition of hydrogen to alkenes

Certain metals such as platinum, palladium and nickel have the ability to adsorb hydrogen. The resulting metal–hydrogen complex can then adsorb an alkene and transfer a pair of hydrogen atoms to the double bond. The product, an alkane, is not strongly held to the metal and desorbs. Because the hydrogen atoms are effectively transferred to the alkene *from the surface of the metal* it is easy to see why they both add to the same face of the alkene (that is, catalytic hydrogenation is a *syn* addition).

e.g.

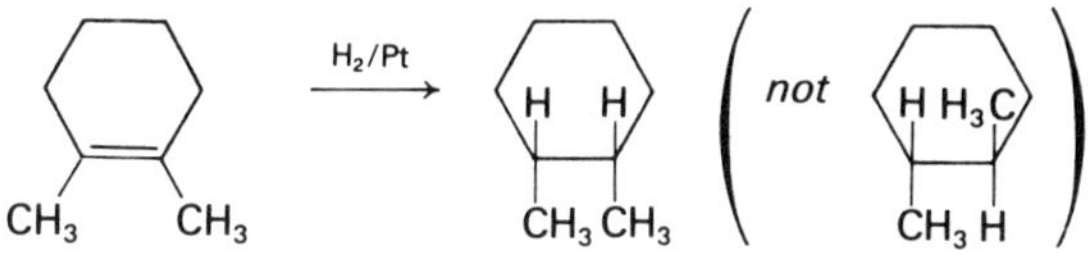

CH_3 CH_3 $\xrightarrow{H_2/Pt}$ H H CH_3 CH_3 (*not* H H_3C CH_3 H)

(c) Addition of hydrogen to alkynes

When an alkyne is hydrogenated to an alkane, stereochemistry cannot come into it:

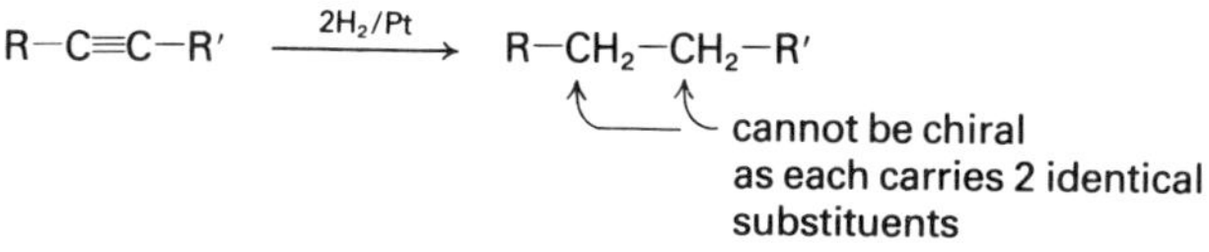

$$R{-}C{\equiv}C{-}R' \xrightarrow{2H_2/Pt} R{-}CH_2{-}CH_2{-}R'$$

cannot be chiral as each carries 2 identical substituents

However, certain catalysts due to Lindlar (based on palladium charcoal) and Brown (based on nickel boride) achieve hydrogenation only as far as the alkene stage and then in theory we may end up with either a *cis* or a *trans* isomer. In practice, it turns out that this kind of catalytic hydrogenation is also a *syn* process. Therefore, for example:

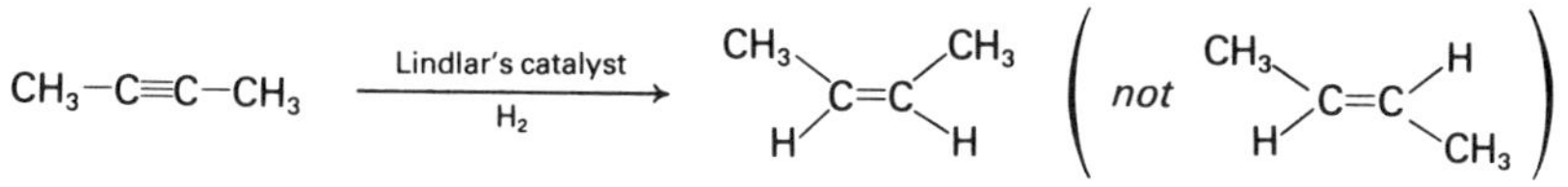

$CH_3{-}C{\equiv}C{-}CH_3$ $\xrightarrow[H_2]{\text{Lindlar's catalyst}}$ CH_3, CH_3 C=C H, H (*not* CH_3, H C=C H, CH_3)

(d) Hydroboration of alkenes

We have already discussed the *anti*-Markovnikov nature of this addition reaction [section 1.7(c)]. Thus, for example, in 1-methylcyclopentene we know which end of the double bond will get the $-BH_2$ group and which end the $-H$ atom. However, we also need to consider the stereochemistry of the addition process which experiment shows to be *syn*. The $-BH_2$ and the $-H$ get transferred at more or less the same time to each end of the **same** side of the original double bond:

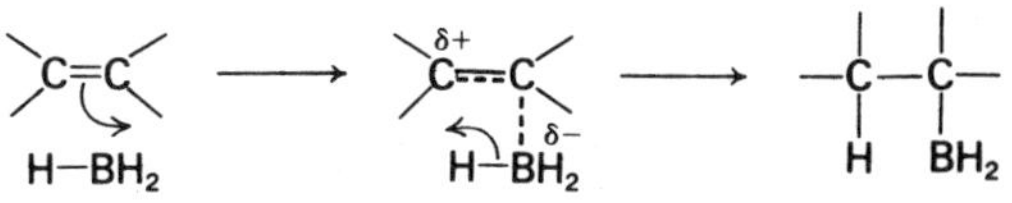

C=C $H{-}BH_2$ ⟶ $\delta+$ C---C $\delta-$ $H{-}BH_2$ ⟶ —C—C— H BH_2

Thus, in the hydroboration of 1-methylcyclopentene:

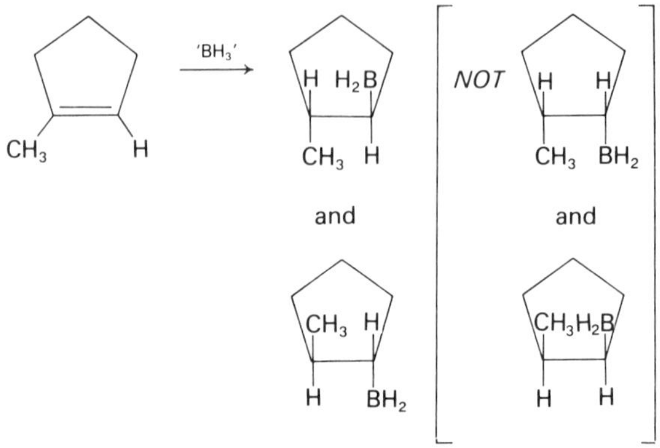

'BH_3'

CH_3 H

H H_2B / CH_3 H and CH_3 H / H BH_2

NOT [H H / CH_3 BH_2 and CH_3H_2B / H H]

In general, the $-BH_2$ group in the product can react in the same way with two more molecules of the alkene, but I will not attempt to draw the outcome in this case! The resulting trialkylboron can be converted to the alcohol by treatment with alkaline hydrogen peroxide and this does not affect the stereochemistry. In this case, therefore, the product is a racemic mixture of:

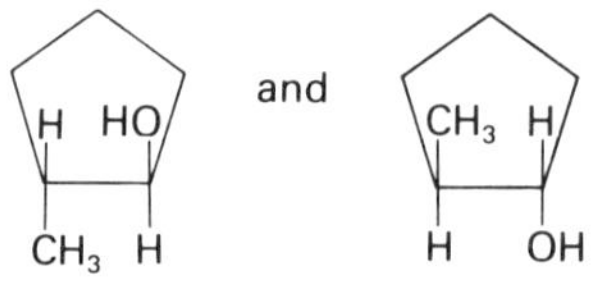

H HO / CH_3 H and CH_3 H / H OH

11.8 *Anti* additions to alkenes and alkynes

(a) Halogenation

It has been established that the halogenation of an alkene involves a cyclic halonium ion that is opened by nucleophilic attack of a halide ion *from the rear* (as in an S_N2 reaction) at either carbon of the cyclic halonium ion. This accounts for the fact that the two halogen atoms become attached on different sides of the double bond, (i.e. an *anti* addition):

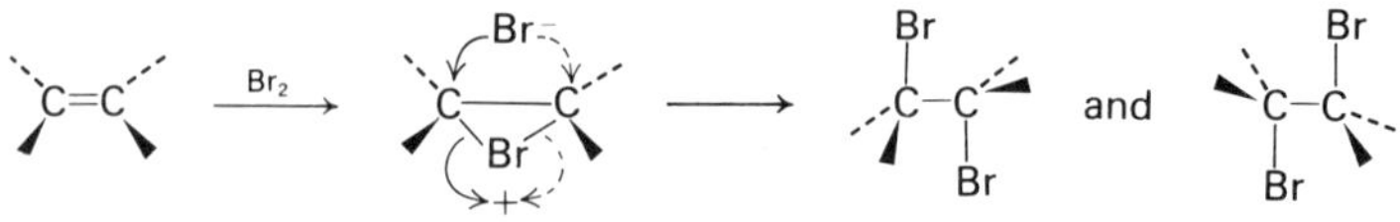

C=C $\xrightarrow{Br_2}$ Br⁻ C—C Br⁺ ⟶ Br C—C Br and Br C—C Br

How do we *know* this is what happens? By studying the stereochemistry of the products. For example, *cis*-2-butene leads to a racemic mixture of products; the meso isomer of 2,3-dibromobutane is *not* formed:

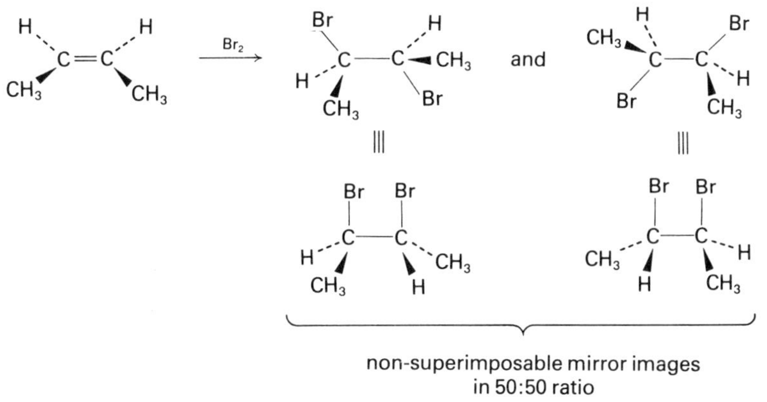

non-superimposable mirror images in 50:50 ratio

(To decide whether a compound such as these is meso, draw it **eclipsed** as I have done here—then a plane of symmetry is obviously either present or, as in this case, absent.) The stereoselective nature of the reaction implies that the central C—C bond is not free to rotate until after both bromines have become attached; this is consistent with the involvement of the cyclic bromonium ion but not of a simple carbocation. Thus, compare:

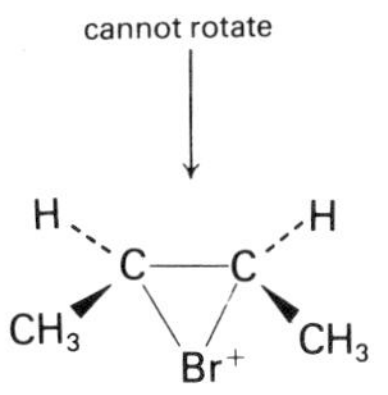

cannot rotate

with:

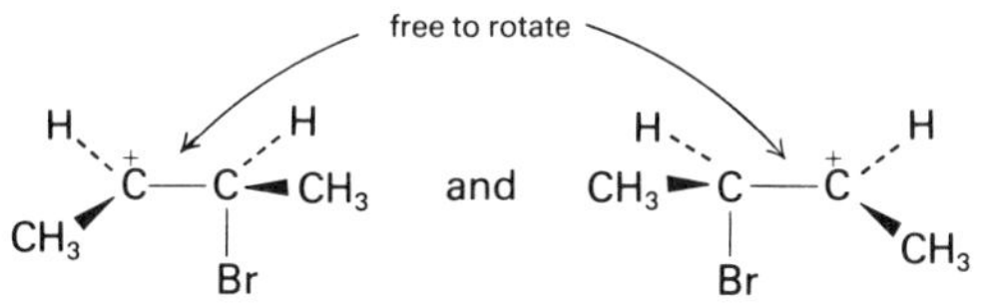

free to rotate

(Attack of Br^- on these non-cyclic species would lead non-selectively to all three possible stereoisomers of 2,3-dibromobutane.)

As a complement to what happens with the *cis* isomer, *trans*-2-butene gives only meso-2,3-dibromobutane (and you should now draw out the mechanism for this reaction and convince yourself you can account for the meso compound as the sole product).

(b) Epoxidation and ring opening

This sequence of reactions leads to *anti* dihydroxylation of an alkene, in contrast to the reaction with potassium permanganate [section 11.7(a)]. The mechanism involves as a first step:

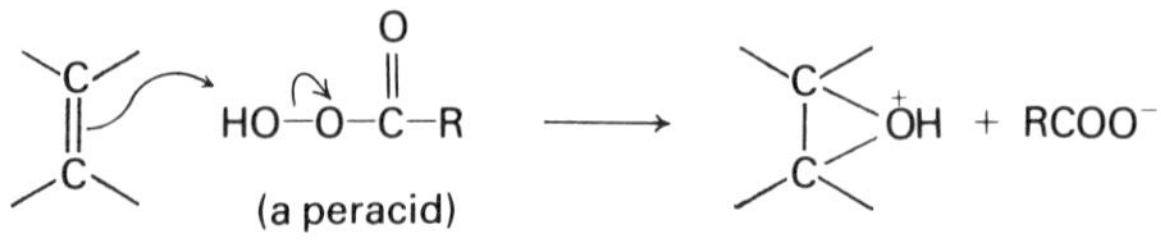

$$\rangle C{=}C\langle \; + \; HO{-}O{-}\overset{O}{\overset{\|}{C}}{-}R \;\longrightarrow\; \text{(protonated epoxide, } \overset{+}{O}H\text{)} \; + \; RCOO^-$$

(a peracid)

Connection Note how similar this is to the first step of bromination. In both cases a positive fragment (Br^+ or HO^+) is transferred to the alkene to give a three membered cyclic compound (bromonium ion or protonated epoxide) and a leaving group (Br^- or $RCOO^-$).

The protonated epoxide can be nucleophilically attacked by water in exactly the same way as the cyclic bromonium ion above was attacked by Br^-:

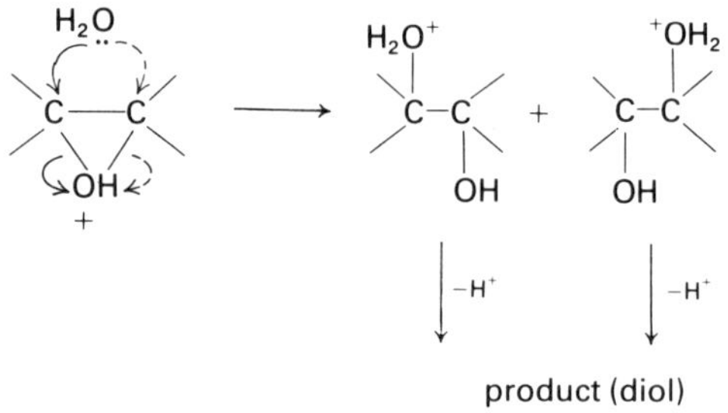

H_2O attacks C—C / $\overset{+}{O}H$ $\longrightarrow$ H_2O^+—C—C—OH + $^+OH_2$—C—C—OH

$-H^+$ $-H^+$

product (diol)

Alternatively, under alkaline conditions the unprotonated epoxide can be opened in the same way by attack of OH^-:

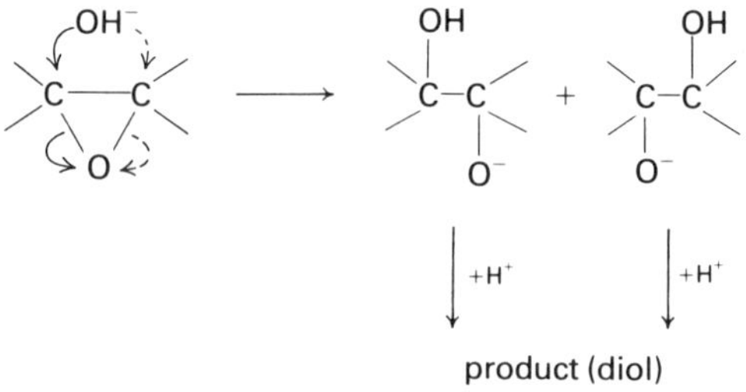

OH^- attacks C—C / O $\longrightarrow$ OH—C—C—O^- + OH—C—C—O^-

$+H^+$ $+H^+$

product (diol)

Again, good evidence for the operation of this mechanism comes from examining the stereochemistry of the products. For example, *trans*-2-butene gives solely meso-2,3-butanediol when subjected to this reaction sequence:

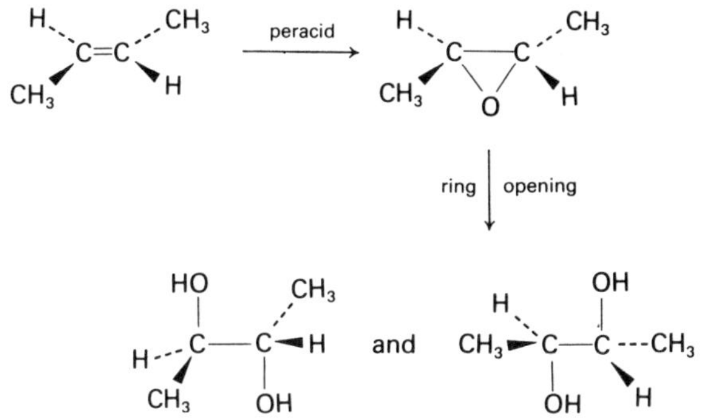

peracid

ring opening

and

Both structures drawn are the same thing and equivalent to:

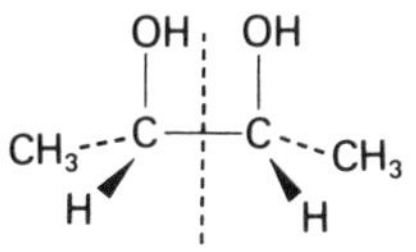

has a plane of symmetry, (i.e. a meso compound)

On the other hand, *cis*-2-butene by the same process gives a racemic mixture of the two optically active isomers and none of the meso isomer, and again you should convince yourself that the *anti* mechanism accounts for this. (Note that the products in cases such as these cannot be distinguished simply by testing the optical activity. The meso compound is optically inactive, and the racemic mixture also has a net rotation of zero. To prove which is which the racemic mixture would have to be **resolved**.)

(c) Summary of *syn* and *anti* additions to *cis*- and *trans*-2-butene

Table 11.1

Alkene	Reaction	Stereochemistry of reaction mechanism	Stereochemistry of product
cis-2-butene	$KMnO_4$	*syn*	meso
	Epoxidation/opening	*anti*	racemic mixture
	Br_2	*anti*	racemic mixture
trans-2-butene	$KMnO_4$	*syn*	racemic mixture
	Epoxidation/opening	*anti*	meso
	Br_2	*anti*	meso

What you should *learn* here is the stereochemical course of each reaction mechanism; then you can always *work out* what the product will be in any given case.

(d) Addition of hydrogen to alkynes

As we have seen [section 11.7(c)] catalytic hydrogenation of alkynes to alkenes is a *syn* addition. The alternative *anti* mode of addition is brought about by sodium in liquid ammonia.

e.g.

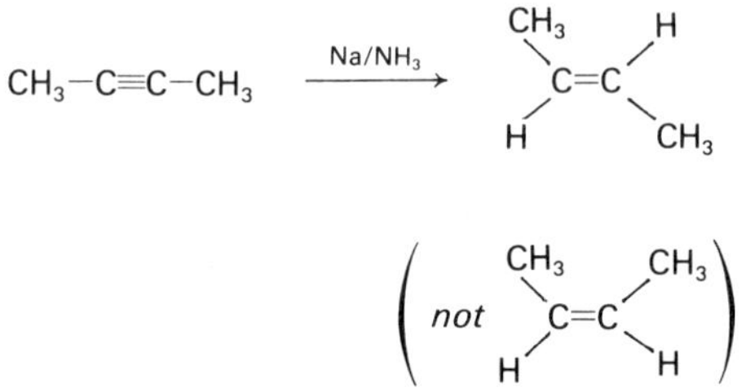

$$CH_3-C\equiv C-CH_3 \xrightarrow{Na/NH_3} \begin{matrix} CH_3 & & H \\ & C=C & \\ H & & CH_3 \end{matrix}$$

$$\left(not \quad \begin{matrix} CH_3 & & CH_3 \\ & C=C & \\ H & & H \end{matrix} \right)$$

The reason for the *anti* mode of addition in this case is not yet fully understood.

11.9 A reminder

Remember that the reactions we have looked at in this chapter by no means always have stereochemical implications. Very often the product would necessarily be the same whether the mechanism was *syn*, *anti*, or whatever.

e.g.

(a) $CH_3-C\equiv C-H \longrightarrow CH_3-CH=CH_2$ — Same product whether Lindlar's catalyst or Na/liq. NH_3 was used.

(b) $(CH_3)_2C=CHCH_3 \longrightarrow (CH_3)_2\underset{OH}{C}-\underset{OH}{CHCH_3}$ — Whether this is a *syn* or *anti* dihydroxylation is irrelevant

(c) $CH_3CH_2Br \xrightarrow[S_N2]{OH^-} CH_3CH_2OH$ — We believe this process 'goes with inversion' but since the compounds are achiral, who knows?!

Index